MATLAB:
A Practical Introduction to
Programming and Problem Solving

MATLAB:
A Practical Introduction to Programming and Problem Solving

By
Stormy Attaway
College of Engineering, Boston University
Boston, MA

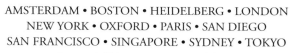

AMSTERDAM • BOSTON • HEIDELBERG • LONDON
NEW YORK • OXFORD • PARIS • SAN DIEGO
SAN FRANCISCO • SINGAPORE • SYDNEY • TOKYO

Butterworth-Heinemann is an imprint of Elsevier

Butterworth-Heinemann is an imprint of Elsevier
30 Corporate Drive, Suite 400, Burlington, MA 01803, USA
Linacre House, Jordan Hill, Oxford OX2 8DP, UK

Library of Congress Cataloging-in-Publication Data
Application submitted

British Library Cataloguing-in-Publication Data
A catalogue record for this book is available from the British Library.

ISBN: 978-0-7506-8762-1

For information on all Butterworth-Heinemann publications
visit our Web site at www.elsevierdirect.com

ATTRIBUTION
MATLAB® and Handle Graphics® are registered trademarks
of The MathWorks, Inc.

Printed in Canada
10 11 12   9 8 7 6 5 4 3

Working together to grow
libraries in developing countries

www.elsevier.com | www.bookaid.org | www.sabre.org

ELSEVIER   BOOK AID
            International   Sabre Foundation

Dedication

This book is dedicated to my two mentors at Boston University: the late Bill Henneman of the Computer Science Department, and the late Merrill Ebner of the Department of Manufacturing Engineering.

Stormy Attaway

Contents

Preface

MOTIVATION

The purpose of this book is to teach basic programming concepts and skills needed for basic problem solving, all using MATLAB® as the vehicle. MATLAB is a powerful software package that has built-in functions to accomplish a diverse range of tasks, from mathematical operations to three-dimensional imaging. Additionally, MATLAB has a complete set of programming constructs that allows users to customize programs to their own specifications.

There are many books that introduce MATLAB. There are two basic flavors of these books: those that demonstrate the use of the built-in functions in MATLAB, with a chapter or two on some programming concepts, and those that cover only the programming constructs without mentioning many of the built-in functions that make MATLAB efficient to use. Someone who learns just the built-in functions will be well-prepared to use MATLAB, but would not understand basic programming concepts. That person then would not be able to learn a language such as C++ or Java without taking another introductory course, or reading another book, on the programming concepts. Conversely, anyone who learns only programming concepts first (using any language) would tend to write highly inefficient code using control statements to solve problems, not realizing that in many cases these are not necessary in MATLAB.

This book instead takes a hybrid approach, introducing both the programming and the efficient uses. The challenge for students is that it is nearly impossible to predict whether they will in fact need to know programming concepts later on or whether a software package such as MATLAB will suffice for their careers. Therefore, the best approach for beginning students is to give them both—the programming concepts and the efficient built-in functions. Since MATLAB is very easy to use, it is a perfect platform for this approach to teaching programming and problem solving.

Since programming concepts are critically important to this book, emphasis is not placed on the time-saving features that evolve with every new MATLAB release. For example, in current versions of MATLAB statistics on variables are available readily in the Workspace window. This is not shown in any detail in the book, since whether this feature is available depends on the version of the software, and because of the desire to explain the concepts in the book.

KEY FEATURES

Side-by-Side Programming Concepts and Built-In Functions

The most important, and unique, feature of this book is that it teaches programming concepts and the use of the built-in functions in MATLAB side by side. It starts with basic programming concepts such as variables, assignments, input/output, selection, and loop statements. Then, throughout the rest of the book, often a problem will be introduced and then solved using the "programming concept" and also using the "efficient method."

Systematic Approach

Another key feature is that the book takes a very systematic, step-by-step approach, building on concepts throughout the book. It is very tempting in a MATLAB text to show built-in functions or features early on with a note that says "we'll do this later." This book does not do that; all functions are covered before they are used in examples. Additionally, basic programming concepts will be explained carefully and systematically. Very basic concepts such as looping to calculate a sum, counting in a conditional loop, and error-checking are not found in many texts but will be covered here.

File Input/Output

Many applications in engineering and the sciences involve manipulating large data sets that are stored in external files. Most MATLAB texts at least mention the **save** and **load** functions, and in some cases also some of the lower level file input/output functions. Since file input and output is so fundamental to so many applications, this book will cover several low-level file input/output functions, as well as reading from and writing to spreadsheet files. Later chapters will also deal with audio and image files. These file input/output concepts are introduced gradually: first **load** and **save** in Chapter 2, then lower level functions in Chapter 8, and finally sound and images in Chapter 13.

User-Defined Functions

User-defined functions are a very important programming concept, and yet many times the nuances and differences between types of functions, function calls versus function headers, and so on, can be very confusing to beginning programmers. Therefore these concepts are introduced gradually. First, arguably the easiest type of functions to understand, those that calculate and return one single value, are demonstrated in Chapter 2. Later, functions that return no values and functions that return multiple values are introduced in Chapter 5. Finally, advanced function features are shown in Chapter 9.

Advanced Programming Concepts

In addition to the basics, some advanced programming concepts such as string manipulation, data structures, recursion, anonymous functions, and variable number of arguments to functions are covered. Sorting, searching, and indexing are also addressed. All these are again approached systematically; for example, cell arrays are covered before they are used in file input functions and as labels on pie charts.

Problem-Solving Tools

In addition to the programming concepts, some basic mathematics necessary for solving many problems will be introduced. These will include statistical functions, solving sets of linear algebraic equations, and fitting curves to data. The use of complex numbers and some calculus (integration and differentiation) will also be introduced. The basic math will be explained and the built-in functions in MATLAB to perform these tasks will be described.

Plots, Imaging, and GUIs

Simple two-dimensional plots are introduced very early in the book (Chapter 2) so that plot examples can be used throughout. Chapter 10 shows more plot types, and demonstrates customizing plots and how the graphics properties are handled in MATLAB. This chapter makes use of strings and cell arrays to customize labels. Also, there is an introduction to image processing and the basics necessary to understand programming Graphical User Interfaces (GUIs) in Chapter 13.

LAYOUT OF TEXT

This text is divided into two parts: Part I covers the programming constructs and demonstrates the programming versus efficient use of built-in functions to solve problems. Part II covers some applications, including plotting, image processing, and mathematics needed in basic problem solving. The first four chapters cover the very basics in MATLAB and in programming, and are all prerequisites for the rest of the book. After that, many chapters in the applications section can be introduced when desired, to produce a customized flow of topics in the book. This is true to an extent, although the order of the chapters has been chosen carefully to ensure that the coverage is systematic.

PART I: PROGRAMMING AND PROBLEM SOLVING USING MATLAB

Chapter 1: Introduction to MATLAB covers expressions, operators, characters, variables, and assignment statements. Scalars, vectors, and matrices are introduced, as are many built-in functions that manipulate them.

Chapter 2: Introduction to MATLAB Programming introduces the idea of algorithms and scripts. This includes simple input and output, and commenting. Scripts are then used to create and customize simple plots, and to do file input and output. Finally, the concept of a user-defined function is introduced with just the type of function that calculates and returns a single value.

Chapter 3: Selection Statements introduces relational expressions and their use in **if** statements, with **else** and **elseif** clauses. The **switch** statement also is demonstrated, as is the concept of choosing from a menu. Also, functions that return logical true or false are introduced.

Chapter 4: Looping introduces the concepts of counted and conditional loops. Many common applications such as summing, counting, and error-checking are covered. Nested loops and combining loops and selection statements are systematically introduced. In this chapter, the idea of vectorizing code is introduced, which essentially means rewriting code that uses loops to more efficiently make use of built-in functions, and the fact that operations can be done on vectors and matrices in MATLAB.

Concepts explained in the first four chapters are assumed throughout the rest of the book.

Chapter 5: MATLAB Programs covers more on scripts and user-defined functions. User-defined functions that return more than one value as well as those that do not return anything are introduced. The concept of a program in MATLAB that consists of a script that calls user-defined functions is demonstrated with examples. A longer menu-driven program is shown as a reference, but could be omitted. Subfunctions and scope of variables are also introduced, as are some debugging techniques.

The concept of a program is used throughout the rest of the book.

Chapter 6: String Manipulation covers many built-in string manipulation functions as well as converting between string and number types. Several examples include using custom strings in plot labels and input prompts.

Chapter 7: Data Structures introduces two main data structures: cell arrays and structures. Once basic structures are covered, more complicated data structures such as nested structures and vectors of structures also are introduced. Cell arrays are used in several applications in later chapters, such as file input in Chapter 8, variable number of function arguments in Chapter 9, and plot labels in Chapter 10, and therefore are considered important and are covered first.

Chapter 8: Advanced File Input and Output covers lower level file input/output statements that require opening and closing the file. Functions that can read the entire file at once as well as those that require reading one line at a time are demonstrated. Additionally, reading from and writing to spreadsheet files and also .mat files that store MATLAB variables are introduced. Cell arrays and string functions are used extensively in this chapter.

Chapter 9: Advanced Functions covers more advanced features and types of functions, such as anonymous functions, nested functions, and recursive functions. Function handles, and their use both with anonymous functions and function functions are introduced. The concept of having a variable number of input and output arguments to a function is introduced; this is implemented using cell arrays. String functions also are used in several examples in this chapter.

PART II: APPLICATIONS

Chapter 10: MATLAB Plots continues with more on the plot functions introduced in Chapter 2. Different plot types such as pie charts are introduced, as are customizing plots using cell arrays and string functions. The notion of handle graphics is covered, and some graphics properties such as line width and color are introduced. Applications that involve reading data from files and then plotting use both cell arrays and string functions.

Chapter 11: Solving Systems of Linear Algebraic Equations introduces the two basic methods that can be used in MATLAB to solve these systems of equations: using a matrix representation, and using symbolic mathematics. First, matrix and vector operations and matrix definitions are described. This section can be covered at any point after Chapter 4. Then, matrix solutions using the Gauss-Jordan and Gauss-Jordan elimination methods are described. Finally, some Symbolic Math Toolbox functions are shown including those that solve equations. This method returns a structure as a result.

Chapter 12: Basic Statistics, Searching, and Sorting starts with some of the built-in statistical and set operations in MATLAB. Since some of these require a sorted data set, methods of sorting are described. Finally, the concepts of indexing into a vector and searching a vector are introduced. Sorting a vector of structures and indexing into a vector of structures are described, but these sections can be omitted if desired. A recursive binary search function can also be omitted.

Chapter 13: Sights and Sounds builds on some handle graphics material from Chapter 10. It also briefly discusses sound files, and introduces image processing. An introduction to programming GUIs is also given. Nested functions are used in the GUI examples. A patch function example uses a structure.

Chapter 14: Advanced Mathematics covers three basic topics: curve fitting, complex numbers, and integration and differentiation in calculus. Symbolic expressions are used in several sections in this chapter.

PEDAGOGICAL FEATURES

There are several pedagogical tools that are used throughout this book that are intended to make it easier to learn.

First, the book takes a conversational tone with sections called "Quick Question!" These are designed to stimulate thought about the material that has just been covered. The question is posed, and then the answer is given. It will be most beneficial if you think about the question before reading the answer! In any case, they should not be skipped over because the answers often contain very useful information.

"Practice!" problems are given throughout the chapters. These are very simple problems that drill the material just covered.

When some problems are introduced, they are solved using both "The Programming Concept" and "The Efficient Method." This facilitates understanding the efficient ways of using MATLAB, and also the programming concepts used in these efficient functions and operators.

The End of Chapter Summary contains, where applicable, several sections:

- **Common Pitfalls**: A list of common mistakes that are made, and how to avoid them.
- **Programming Style Guidelines**: In order to encourage "good" programs, which others can actually understand, the programming chapters will have guidelines that will make programs easier to read and understand and therefore easier to work with and to modify.
- **Key Terms**: A list of the key terms covered in the chapter, in sequence.
- **MATLAB Functions and Commands**: A list of the MATLAB built-in functions and commands covered in the chapter, in the order covered. Throughout the text, these are given in bold type.
- **MATLAB Operators**: A list of the MATLAB operators covered in the chapter, in the order covered.
- **Exercises**: A comprehensive set of exercises, ranging from the rote to more engaging applications.

ADDITIONAL BOOK RESOURCES

A companion web site with additional teaching resources is available for faculty using this book as a text for their course(s). Please visit www.textbooks. elsevier.com/9780750687621 to register for access to:

- Instructor solutions manual for end-of-chapter problems
- Electronic figures from the text for creation of lecture slides
- Downloadable M-files for all examples in the text

Other book-related resources will also be posted there from time to time.

ACKNOWLEDGMENTS

I am indebted to many, many family members, colleagues, mentors, and students.

Throughout the last 22 years of coordinating and teaching the basic computation courses for the College of Engineering at Boston University, I have been blessed with many fabulous students as well as graduate teaching fellows and undergraduate teaching assistants (TAs). There have been hundreds of TAs over the years, too many to name individually, but I thank them all for their support. In particular the following TAs have been very helpful in the last year reviewing drafts of this text and suggesting examples: Edy Tan, Megan Smith, Brandon Phillips, Carly Sherwood, Ashmita Randhawa, Mike Green, Kevin Ryan, and Brian Hsu. Kevin Ryan created the script to produce the cover illustration.

A number of colleagues have been very encouraging throughout the years. In particular, I would like to thank Tom Bifano for his support and for his GUI example suggestions.

I would like to thank all the reviewers of the proposal and drafts of this book. Their comments have been extremely helpful and I hope I have incorporated their suggestions to their satisfaction. They include Mark Cawood, Clemson University; Dieter Schmidt, University of Cincinnati; Matthew Verleger, Purdue University; Peter Orono, Indiana University–Purdue University Indianapolis; James Kang, California State Polytechnic University–Pomona; David Rockstraw, New Mexico State University; and James Holloway, University of Michigan.

Also, I thank those at Elsevier who helped to make this book possible including Tim Pitts, Joseph Hayton, Publisher; Maria Alonso, Assistant Editor; Christie Jozwiak, Production Manager; and Eric Decicco, Cover Designer/Illustrator.

Finally, thanks go to all my family. My parents encouraged me at an early age to read and to write. Thanks also to my husband Ted de Winter of Boston

University's Department of Mechanical Engineering for his encouragement and good-natured taking care of the Saturday shopping and gardening while I worked on this project!

The photo of Ted fishing in the image processing section was taken by Wes Karger.

PART

1

Programming and Problem Solving Using MATLAB

CHAPTER 1

Introduction to MATLAB

KEY WORDS

prompt	nesting	row vector
programs	call a function	column vector
script files	arguments	scalar
variables	returning values	elements
assignment statement	constants	array
assignment operator	types	array operations
user	classes	iterate
initializing	floating point	step value
incrementing	characters	concatenating
decrementing	strings	index
identifier names	type casting	subscript
reserved words	saturation arithmetic	index vector
mnemonic	random numbers	transposing
default	seed	unwinding a matrix
unary	pseudo-random	linear indexing
operand	character encoding	vector of variables
binary	character set	empty vector
precedence	vectors	deleting elements
associativity	matrices	

CONTENTS

MATLAB® is a very powerful software package that has many built-in tools for solving problems and for graphical illustrations. The simplest method for using the MATLAB product is interactively; an expression is entered by the user and MATLAB immediately responds with a result. It is also possible to write programs in MATLAB, which are essentially groups of commands that are executed sequentially.

This chapter will focus on the basics, including many operators and built-in functions that can be used in interactive expressions. Means of storing values, including vectors and matrices, will also be introduced.

1.1 GETTING INTO MATLAB

MATLAB is a mathematical and graphical software package; it has numerical, graphical, and programming capabilities. It has built-in functions to do many operations, and there are toolboxes that can be added to augment these functions (e.g., for signal processing). There are versions available for different hardware platforms, and there are both professional and student editions.

When the MATLAB software is started, a window is opened: the main part is the Command Window (see Figure 1.1). In the Command Window, there is a statement that says:

In the Command Window, you should see:

> \>>

The \>> is called the prompt. In the Student Edition, the prompt appears as:

> EDU\>>

In the Command Window, MATLAB can be used interactively. At the prompt, any MATLAB command or expression can be entered, and MATLAB will immediately respond with the result.

It is also possible to write *programs* in MATLAB, which are contained in *script files* or M-files. Programs will be introduced in Chapter 2.

There are several commands that can serve as an introduction to MATLAB and allow you to get help:

- **info** will display contact information for the product

- **demo** has demos of several options in MATLAB

- **help** will explain any command; **help help** will explain how help works

- **helpbrowser** opens a Help Window

FIGURE 1.1

MATLAB Command Window.

- **lookfor** searches through the help for a specific string (be aware that this can take a long time)

To get out of MATLAB, either type **quit** at the prompt, or choose File, then Exit MATLAB from the menu.

In addition to the Command Window, there are several other windows that can be opened and may be opened by default. What is described here is the default layout for these windows, although there are other possible configurations. Directly above the Command Window, there is a pull-down menu for the Current Directory. The folder that is set as the Current Directory is where files will be saved. By default, this is the Work Directory, but that can be changed.

To the left of the Command Window, there are two tabs for Current Directory Window and Workspace Window. If the Current Directory tab is chosen, the files stored in that directory are displayed. The Command History Window shows commands that have been entered, not just in the current session (in the current Command Window), but previously as well. This default configuration can be altered by clicking Desktop, or using the icons at the top-right corner of each window: either an "x," which will close that particular window; or a curled arrow, which in its initial state pointing to the upper right lets you undock that window. Once undocked, clicking the curled arrow pointing to the lower right will dock the window again.

1.2 VARIABLES AND ASSIGNMENT STATEMENTS

In order to store a value in a MATLAB session, or in a program, a *variable* is used. The Workspace Window shows variables that have been created. One easy way to create a variable is to use an *assignment statement*. The format of an assignment statement is

```
variablename = expression
```

The variable is always on the left, followed by the *assignment operator*, = (unlike in mathematics, the single equal sign does *not* mean equality), followed by an expression. The expression is evaluated and then that value is stored in the variable. For example, this is the way it would appear in the Command Window:

```
>> mynum = 6
mynum =
    6
>>
```

Here, the *user* (the person working in MATLAB) typed mynum = 6 at the prompt, and MATLAB stored the integer 6 in the variable called *mynum*, and then displayed the result followed by the prompt again. Since the equal sign is the assignment operator, and does not mean equality, the statement should be read as "mynum gets the value of 6" (*not* "mynum equals 6").

Note that the variable name must always be on the left, and the expression on the right. An error will occur if these are reversed.

```
>> 6 = mynum
??? 6 = mynum
     |
Error: The expression to the left of the equals sign is not
   a valid target for an assignment.
>>
```

Putting a semicolon at the end of a statement suppresses the output. For example,

```
>> res = 9 - 2;
>>
```

This would assign the result of the expression on the right side, the value 7, to the variable *res*; it just doesn't show that result. Instead, another prompt appears immediately. However, at this point in the Workspace Window the variables *mynum* and *res* can be seen.

Note: In the remainder of the text, the prompt that appears after the result will not be shown.

The spaces in a statement or expression do not affect the result, but make it easier to read. The following statement that has no spaces would accomplish exactly the same thing as the previous statement:

```
>> res = 9-2;
```

MATLAB uses a default variable named *ans* if an expression is typed at the prompt and it is not assigned to a variable. For example, the result of the expression 6 + 3 is stored in the variable *ans*:

```
>> 6 + 3
ans =
      9
```

This default variable is reused any time just an expression is typed at the prompt.

A short-cut for retyping commands is to press the up-arrow ↑, which will go back to the previously typed command(s). For example, if you decided to assign the result of the expression 6 + 3 to the variable *res* instead of using the default *ans*, you could press the up-arrow and then the left-arrow to modify the command rather than retyping the whole statement:

```
>> res = 6 + 3
res =
      9
```

This is very useful, especially if a long expression is entered with an error, and you want to go back to correct it.

To change a variable, another assignment statement can be used that assigns the value of a different expression to it. Consider, for example, the following sequence of statements:

```
>> mynum = 3
mynum =
      3
>> mynum = 4 + 2
mynum =
      6
```

```
>> mynum = mynum + 1
mynum =
        7
```

In the first assignment statement, the value 3 is assigned to the variable *mynum*. In the next assignment statement, *mynum* is changed to have the value of the expression 4 + 2, or 6. In the third assignment statement, *mynum* is changed again, to the result of the expression mynum + 1. Since at that time *mynum* had the value 6, the value of the expression was 6 + 1, or 7.

At that point, if the expression mynum + 3 is entered, the default variable *ans* is used since the result of this expression is not assigned to a variable. Thus, the value of *ans* becomes 10 but *mynum* is unchanged (it is still 7). Note that just typing the name of a variable will display its value.

```
>> mynum + 3
ans =
        10
>> mynum
mynum =
        7
```

1.2.1 Initializing, Incrementing, and Decrementing

Frequently, values of variables change. Putting the first or initial value in a variable is called *initializing* the variable.

Adding to a variable is called *incrementing*. For example, the statement

```
mynum = mynum + 1
```

increments the variable *mynum* by 1.

QUICK QUESTION!

How can 1 be subtracted from the value of a variable called *num*?

Answer:

num = num − 1;
This is called *decrementing* the variable.

1.2.2 Variable Names

Variable names are an example of *identifier names*. We will see other examples of identifier names, such as filenames, in future chapters. The rules for identifier names are:

- The name must begin with a letter of the alphabet. After that, the name can contain letters, digits, and the underscore character (e.g., value_1), but it cannot have a space.

- There is a limit to the length of the name; the built-in function **namelengthmax** tells how many characters this is.

- MATLAB is case-sensitive. That means that there is a difference between upper- and lowercase letters. So, variables called *mynum*, *MYNUM*, and *Mynum* are all different.

- There are certain words called *reserved words* that cannot be used as variable names.

- Names of built-in functions can, but should not, be used as variable names.

Additionally, variable names should always be ***mnemonic***, which means they should make some sense. For example, if the variable is storing the radius of a circle, a name such as "radius" would make sense; "x" probably wouldn't.

The Workspace Window shows the variables that have been created in the current Command Window and their values.

The following commands relate to variables:

- **who** shows variables that have been defined in this Command Window (this just shows the names of the variables)

- **whos** shows variables that have been defined in this Command Window (this shows more information on the variables, similar to what is in the Workspace Window)

- **clear** clears out all variables so they no longer exist

- **clear** *variablename* clears out a particular variable

If nothing appears when **who** or **whos** is entered, that means there aren't any variables! For example, in the beginning of a MATLAB session, variables could be created and then selectively cleared (remember that the semicolon suppresses output):

```
>> who
>> mynum = 3;
>> mynum + 5;

>> who
Your variables are:
ans  mynum

>> clear mynum
```

```
>> who
Your variables are:
ans
```

1.3 EXPRESSIONS

Expressions can be created using values, variables that have already been created, operators, built-in functions, and parentheses. For numbers, these can include operators such as multiplication, and functions such as trigonometric functions. An example of such an expression would be:

```
>> 2 * sin(1.4)
ans =
      1.9709
```

1.3.1 The Format Function and Ellipsis

The *default* in MATLAB is to display numbers that have decimal places with four decimal places, as already shown. The **format** command can be used to specify the output format of expressions. There are many options, including making the format **short** (the default) or **long**. For example, changing the format to **long** will result in 15 decimal places. This will remain in effect until the format is changed back to **short**, as demonstrated with an expression and with the built-in value for **pi**.

```
>> format long
>> 2 * sin(1.4)
ans =
      1.970899459976920

>> pi
ans =
      3.141592653589793

>> format short
>> 2 * sin(1.4)
ans =
      1.9709

>> pi
ans =
      3.1416
```

The **format** command can also be used to control the spacing between the MATLAB command or expression and the result; it can be either **loose** (the default) or **compact**.

```
>> format loose
>> 2^7

ans =

    128

>> format compact
>> 2^7
ans =
    128
```

Especially long expressions can be continued on the next line by typing three (or more) periods, which is the continuation operator, or the **ellipsis**. For example,

```
>> 3 + 55 - 62 + 4 - 5 ...
   + 22 - 1

ans =
    16
```

1.3.2 Operators

There are in general two kinds of operators: *unary* operators, which operate on a single value or *operand*; and *binary* operators, which operate on two values or operands. The symbol "−", for example, is both the unary operator for negation and the binary operator for subtraction.

Here are some of the common operators that can be used with numeric expressions:

```
+       addition
−       negation, subtraction
*       multiplication
/       division (divided by e.g. 10/5 is 2)
\       division (divided into e.g. 5\10 is 2)
^       exponentiation (e.g., 5^2 is 25)
```

1.3.2.1 Operator Precedence Rules

Some operators have *precedence* over others. For example, in the expression 4 + 5 * 3, the multiplication takes precedence over the addition, so first 5 is multiplied by 3, then 4 is added to the result. Using parentheses can change the precedence in an expression:

```
>> 4 + 5 * 3
ans =
    19
>> (4 + 5) * 3
ans =
    27
```

Within a given precedence level, the expressions are evaluated from left to right (this is called the *associativity*).

Nested parentheses are parentheses inside of others; the expression in the *inner parentheses* is evaluated first. For example, in the expression 5 - (6 * (4 + 2)), first the addition is performed, then the multiplication, and finally the subtraction to result in −31. Parentheses can also be used simply to make an expression clearer. For example, in the expression ((4 + (3 * 5)) - 1) the parentheses are not necessary, but are used to show the order in which the expression will be evaluated.

For the operators that have been covered so far, the following is the precedence (from the highest to the lowest):

( )	parentheses
^	exponentiation
−	negation
*, /, \	all multiplication and division
+, −	addition and subtraction

 PRACTICE 1.1

Think about what the results would be for the following expressions, and then type them to verify your answers:

```
4 ^ 2 - 1
4 ^ (2 - 1)
2\3
4 * 2 - 9/3
5 - - 3
```

1.3.3 Built-In Functions and Help

There are many, many built-in functions in MATLAB. The **help** command can be used to find out what functions MATLAB has, and also how to use them. For example, typing **help** at the prompt in the Command Window will show a list of help topics, which are groups of related functions. This is a very long list; the most elementary help topics are in the beginning.

For example, one of these is listed as **matlab\elfun**; it includes the elementary math functions. Another of the first help topics is **matlab\ops**, which shows the operators that can be used in expressions.

To see a list of the functions contained within a particular help topic, type **help** followed by the name of the topic. For example,

```
>> help elfun
```

will show a list of the elementary math functions. It is a very long list, and is broken into trigonometric (for which the default is radians, but there are equivalent functions that instead use degrees), exponential, complex, and rounding and remainder functions.

To find out what a particular function does and how to call it, type **help** and then the name of the function. For example,

```
>> help sin
```

will give a description of the **sin** function.

To *call* a function, the name of the function is given followed by the *argument(s)* that are passed to the function in parentheses. Most functions then *return* value(s). For example, to find the absolute value of –4, the following expression would be entered:

```
>> abs(-4)
```

which is a *call* to the function **abs**. The number in the parentheses, the –4, is the *argument*. The value 4 would then be *returned* as a result.

QUICK QUESTION!

What would happen if you use the name of a function, for example sin, as a variable name?

Answer:

This is allowed in MATLAB, but then sin could not be used as the built-in function, until the variable is cleared. For example, try the following sequence:

```
>> sin(3.1)
ans =
   0.0416
>> sin = 45
sin =
   45
>> sin(3.1)
```

??? Subscript indices must either be real positive integers or logicals.

```
>> who
Your variables are:
ans sin
>> clear sin
>> who
Your variables are:
ans
>> sin(3.1)
ans =
   0.0416
```

In addition to the trigonometric functions, the elfun help topic also has some rounding and remainder functions that are very useful. Some of these include **fix, floor, ceil, round, rem,** and **sign**.

The **rem** function returns the remainder from a division; for example 5 goes into 13 twice with a remainder of 3, so the result of this expression is 3:

```
>> rem(13,5)
ans =
   3
```

QUICK QUESTION!

What would happen if you reversed the order by mistake, and typed the following:

```
rem(5,13)
```

Answer:

The **rem** function is an example of a function that has two arguments passed to it. The **rem** function divides the second argument into the first. In this case, the second argument, 13, goes into 5 zero times with a remainder of 5, so 5 would be returned as a result.

PRACTICE 1.2

Use the **help** function to find out what the rounding functions **fix**, **floor**, **ceil**, and **round** do. Experiment with them by passing different values to the functions: some negative, some positive, some with fractions less than 0.5 and some greater. *It is very important when testing functions that you thoroughly test by trying different kinds of arguments!*

Another function in the elfun help topic is the **sign** function, which returns 1 if the argument is positive, 0 if it is 0, and –1 if it is negative. For example,

```
>> sign(-5)
ans =
    -1
>> sign(3)
ans =
    1
```

1.3.4 Constants

Variables are used to store values that can change, or that are not known ahead of time. Most languages also have the capacity to store *constants*, which are values that are known ahead of time, and cannot possibly change. An example of a constant value would be **pi**, or π, which is 3.14159…. In MATLAB, there are functions that return some of these constant values. Some of these include:

pi	3.14159….
i	$\sqrt{1}$
j	$\sqrt{1}$
inf	infinity ∞
NaN	stands for "not a number"; e.g., the result of 0/0

QUICK QUESTION!

There is no built-in constant for e (2.718), so how can that value be obtained in MATLAB?

Answer:

Use the exponential function **exp**; e is equivalent to exp(1).

```
>> exp(1)
ans =
    2.7183
```

1.3.5 Types

Every expression, or variable, has a *type* associated with it. MATLAB supports many types of values, which are called *classes*. A class is essentially a combination of a type and the operations that can be performed on values of that type. For example, there are types to store different kinds of numbers.

For float or real numbers, or in other words numbers with a decimal place (e.g., 5.3), there are two basic types: **single** and **double**. The name of the type **double** is short for double precision; it stores larger numbers than **single**. MATLAB uses a *floating point* representation for these numbers.

For integers, there are many integer types (e.g., **int8**, **int16**, **int32**, and **int64**). The numbers in the names represent the number of bits used to store values of that type. For example, the type **int8** uses eight bits altogether to store the integer and its sign. Since one bit is used for the sign, this means that seven bits are used to store the actual number. Each bit stores the number in binary (0's or 1's), and 0 is also a possible value, which means that $2 \wedge 7 - 1$ or 127 is the largest number that can be stored. The range of values that can be stored in **int8** is actually from -128 to $+127$. This range can be found for any type by passing the name of the type as a string (which means in single quotes) to the functions **intmin** and **intmax**. For example,

```
>> intmin('int8')
ans =
   -128
>> intmax('int8')
ans =
    127
```

The larger the number in the type name, the larger the number that can be stored in it. We will for the most part use the type **int32** when an integer type is required.

The type **char** is used to store either single *characters* (e.g., 'x') or *strings*, which are sequences of characters (e.g., 'cat'). Both characters and strings are enclosed in single quotes.

The type **logical** is used to store true/false values.

If any variables have been created in the Command Window, they can be seen in the Workspace Window. In that window, for every variable, the variable name, value, and class (which is essentially its type) can be seen. Other attributes of variables can also be seen in the Workspace Window. Which attributes are visible by default depends on the version of MATLAB. However, when the Workspace Window is chosen, clicking View allows the user to choose which attributes will be displayed.

By default, numbers are stored as the type **double** in MATLAB. There are, however, many functions that convert values from one type to another. The names of these functions are the same as the names of the types just shown. They can be used as functions to convert a value to that type. This is called *casting* the value to a different type, or type casting. For example, to convert a value from the type **double**, which is the default, to the type **int32**, the function **int32** would be used. Typing the following assignment statement:

```
>> val = 6+3
```

would result in the number 9 being stored in the variable *val*, with the default type of **double**, which can be seen in the Workspace Window. Subsequently, the assignment statement

```
>> val = int32(val);
```

would change the type of the variable to **int32**, but would not change its value. If we instead stored the result in another variable, we could see the difference in the types by using **whos**.

```
>> val = 6 + 3;
>> vali = int32(val);
>> whos
```

Name	Size	Bytes	Class	Attributes
val	1x1	8	double	
vali	1x1	4	int32	

One reason for using an integer type for a variable is to save space.

QUICK QUESTION!

What would happen if you go beyond the range for a particular type? For example, the largest integer that can be stored in **int8** is 127, so what would happen if we type cast a larger integer to the type **int8**?

```
>> int8(200)
```

Answer:
The value would be the largest in the range, in this case 127. If we instead use a negative number that is smaller than the

lowest value in the range, its value would be −128. This is an example of what is called **saturation arithmetic**.

```
>> int8(200)
ans =
   127
>> int8(-130)
ans =
  -128
```

PRACTICE 1.3

1. Calculate the range of integers that can be stored in the type **int16**. Use **intmin** and **intmax** to verify your results.

2. Enter an assignment statement and view the type of the variable in the Workspace Window. Then, change its type.

1.3.6 Random Numbers

When a program is being written to work with data, and the data is not yet available, it is often useful to test the program first by initializing the data variables to *random numbers*. There are several built-in functions in MATLAB that *generate* random numbers, some of which will be illustrated in this section.

Random number generators or functions are not truly random. Basically, the way it works is that the process starts with one number, called a *seed*. Frequently, the initial seed is either a predetermined value or it is obtained from the built-in clock in the computer. Then, based on this seed, a process determines the next random number. Using that number as the seed the next time, another random number is generated, and so forth. These are actually called *pseudo-random*; they are not truly random because there is a process that determines the next value each time.

The function **rand** can be used to generate random real numbers; calling it generates one random real number in the range from 0 to 1. There are no arguments passed to the **rand** function. Here are two examples of calling the **rand** function:

```
>> rand
ans =
    0.9501
>> rand
ans =
    0.2311
```

The seed for the **rand** function will always be the same each time MATLAB is started, unless the state is changed, for example, by the following:

```
rand('state',sum(100*clock))
```

This uses the current date and time that are returned from the built-in **clock** function to set the seed. Note: this is done only once in any given MATLAB session to set the seed; the **rand** function can then be used as shown earlier any number of times to generate random numbers.

Since **rand** returns a real number in the range from 0 to 1, multiplying the result by an integer N would return a random real number in the range from 0 to N. For example, multiplying by 10 returns a real in the range from 0 to 10, so this expression

```
rand*10
```

would return a result in the range from 0 to 10.

To generate a random real number in the range from *low* to *high*, first create the variables *low* and *high*. Then, use the expression rand*(high–low) + low. For example, the sequence

```
>> low = 3;
>> high = 5;
>> rand*(high-low)+low
```

would generate a random real number in the range from 3 to 5.

The Programming Concept

In most programming languages, the random function returns a real number, so the real number would then have to be rounded to generate a random integer. For example,

```
>> round(rand*10)
```

would generate one random integer in the range from 0 to 10 (rand*10 would generate a random real in the range from 0 to 10; from that rounding will return an integer). Or, to generate a random integer in a range:

```
>> low = 2;
>> high = 6;
>> round(rand*(high-low)+low)
```

This would generate a random integer in the range from 2 to 6.

The Efficient Method

However, in MATLAB, there is another built-in function that specifically generates random integers, **randint**. Calling the function with **randint(1,1,N)** generates one random integer in the range from 0 to N − 1. The first two arguments essentially specify that one random integer will be returned; the third argument gives the range of that random integer. For example,

```
>> randint(1,1,4)
```

generates a random integer in the range from 0 to 3. Note: Even though this creates random integers, the type is actually the default type **double**.

A range can also be passed to the **randint** function. For example, the following specifies a random integer in the range from 1 to 20:

```
>> randint(1,1,[1,20])
```

 PRACTICE 1.4

Generate a random:

- Real number in the range from 0 to 1
- Real number in the range from 0 to 50
- Real number in the range from 10 to 20
- Integer in the range from 1 to 10

1.4 CHARACTERS AND ENCODING

A character in MATLAB is represented using single quotes (e.g., 'a' or 'x'). The quotes are necessary to denote a character; without them, the letter would be interpreted as a variable name. Characters are put in an order using what is called a *character encoding*. In the character encoding, all characters in the computer's *character set* are placed in a sequence and given equivalent integer values. The character set includes all letters of the alphabet, digits, punctuation marks, and more; basically, all the keys on a keyboard are characters. Special characters such as the Enter key are also included. So, 'x', '!', and '3' are all characters. With quotes, '3' is a character, not a number. The most common character encoding is the American Standard Code for Information Interchange, or ASCII. Standard ASCII has 128 characters, which have equivalent integer values from 0 to 127. The first 32 (integer values 0 through 31) are nonprinting characters. The letters of the alphabet are in order, which means 'a' comes before 'b', then 'c', and so forth.

The numeric functions can be used to convert a character to its equivalent numeric value (e.g., **double** will convert to a double value, **int32** will convert to an integer value using 32 bits, etc.). For example, to convert the character 'a' to its numeric equivalent, the following statement could be used:

```
>> numequiv = double('a')
numequiv =
    97
```

This stores the **double** value 97 in the variable *numequiv*, which shows that the character 'a' is the ninety-eighth character in the character encoding (since the equivalent numbers begin at 0). It doesn't matter which number type is used to convert 'a', for example,

```
>> numequiv = int32('a')
```

would also store the integer value 97 in the variable *numequiv*. The only difference between these will be the type of the resulting variable (**double** in the first case, **int32** in the second).

The function **char** does the reverse; it converts from any number type to the type char:

```
>> char(numequiv)
ans =
a
```

Note that the quotes are not printed.

Since the letters of the alphabet are in order, the character 'b' has the equivalent value of 98, 'c' is 99, and so on. Math can be done on characters. For example, to get the next character in the collating sequence, 1 can be added either to the integer or the character:

```
>> numequiv = double('a');
>> char(numequiv + 1)
ans =
b
>> 'a' + 2
ans =
    99
```

MATLAB also handles *strings*, which are a sequence of characters in single quotes. For example, using the **double** function on a string will show the equivalent numerical value of all characters in the string:

```
>> double('abcd')
ans =
     97    98    99   100
```

To shift the characters of a string *up* in the character encoding, an integer value can be added to a string. For example, the following expression will shift by one:

```
>> char('abcd'+ 1)
ans =
bcde
```

PRACTICE 1.5

1. Find the numerical equivalent of the character 't'.
2. Find the character equivalent of 112.

1.5 VECTORS AND MATRICES

Vectors and *matrices* are used to store sets of values, all of which are the same type. A vector can be either a *row vector* or a *column vector*. A matrix can be visualized as a table of values. The dimensions of a matrix are r × c, where r is the number of rows and c is the number of columns. This is pronounced "r by c." If a vector has *n* elements, a row vector would have the dimensions 1 × *n*, and a column vector would have the dimensions *n* × 1. A *scalar* (one value) has the dimensions 1 × 1. Therefore, vectors and scalars are actually just subsets of matrices. Here are some diagrams showing, from left to right, a scalar, a column vector, a row vector, and a matrix:

5

3
7
4

5	88	3	11

9	6	3
5	7	2
4	33	8

The scalar is 1 × 1, the column vector is 3 × 1 (3 rows by 1 column), the row vector is 1 × 4 (1 row by 4 columns), and the matrix is 3 × 3. All the values stored in these matrices are stored in what are called *elements*.

MATLAB is written to work with matrices; the name MATLAB is short for "matrix laboratory." For this reason, it is very easy to create vector and matrix variables, and there are many operations and functions that can be used on vectors and matrices.

A vector in MATLAB is equivalent to what is called a one-dimensional *array* in other languages. A matrix is equivalent to a two-dimensional array. Usually, even in MATLAB, some operations that can be performed on either vectors or matrices are referred to as *array operations*. The term *array* also frequently is used to mean generically either a vector or a matrix.

1.5.1 Creating Row Vectors

There are several ways to create row vector variables. The most direct way is to put the values that you want in the vector in square brackets, separated by either spaces or commas. For example, both of these assignment statements create the same vector v:

```
>> v = [1  2  3  4]
v =
    1    2    3    4
>> v = [1,2,3,4]
v =
    1    2    3    4
```

Both of these create a row vector variable that has four elements; each value is stored in a separate element in the vector.

1.5.1.1 *The Colon Operator and* Linspace *Function*

If, as in the earlier examples, the values in the vector are regularly spaced, the *colon operator* can be used to *iterate* through these values. For example, 1:5 results in all the integers from 1 to 5:

```
>> vec = 1:5
vec =
    1    2    3    4    5
```

Note that in this case, the brackets [ ] are not necessary to define the vector.

With the colon operator, a *step value* can also be specified with another colon, in the form (first:step:last). For example, to create a vector with all integers from 1 to 9 in steps of 2:

```
>> nv = 1:2:9
nv =
    1    3    5    7    9
```

QUICK QUESTION!

What happens if adding the step value would go beyond the range specified by the last, for example,

```
1:2:6
```

Answer:

This would create a vector containing 1, 3, and 5 and then adding 2 to the 5 would go beyond 6, so the vector stops at 5; the result would be

```
1    3    5
```

QUICK QUESTION!

How can you use the colon operator to generate the following vector?

| 9 | 7 | 5 | 3 | 1 |

Answer:

$$9:-2:1$$

The step operator can be a negative number, so the resulting sequence is in descending order.

Similarly, the **linspace** function creates a linearly spaced vector; **linspace(x,y,n)** creates a vector with n values in the inclusive range from x to y. For example, the following creates a vector with five values linearly spaced between 3 and 15, including the 3 and 15:

```
>> ls = linspace(3,15,5)
ls =
      3    6    9   12   15
```

Vector variables can also be created using existing variables. For example, a new vector is created here consisting first of all the values from *nv* followed by all values from *ls*:

```
>> newvec = [nv ls]
newvec =
     1   3   5   7   9   3   6   9   12   15
```

Putting two vectors together like this to create a new one is called *concatenating* the vectors.

1.5.1.2 Referring to and Modifying Elements

A particular element in a vector is accessed using the name of the vector variable and the element number (or *index*, or *subscript*) in parentheses. In MATLAB, the indices start at 1. Normally, diagrams of vectors and matrices show the indices; for example, for the variable *newvec* created earlier the indices 1–10 of the elements are shown above the vector:

1	2	3	4	5	6	7	8	9	10
1	3	5	7	9	3	6	9	12	15

For example, the fifth element in the vector *newvec* is a 9.

```
>> newvec(5)
ans =
     9
```

A subset of a vector, which would be a vector itself, can also be obtained using the colon operator. For example, the following statement would get the fourth through sixth elements of the vector *newvec*, and store the result in a vector variable *b*:

```
>> b = newvec(4:6)
b =
   7  9  3
```

Any vector can be used for the indices in another vector, not just one created using the colon operator. For example, the following would get the first, fifth, and tenth elements of the vector *newvec*:

```
>> newvec([1 5 10])
ans =
    1  9  15
```

The vector [1 5 10] is called an ***index vector***; it specifies the indices in the original vector that are being referenced.

The value stored in a vector element can be changed by specifying the index or subscript. For example, to change the second element from the vector *b* to now store the value 11 instead of 9:

```
>> b(2) = 11
b =
   7  11  3
```

By using an index, a vector can also be extended. For example, the following creates a vector that has three elements. By then referring to the fourth element in an assignment statement, the vector is extended to have four elements.

```
>> rv = [3 55 11]
rv =
    3  55  11
>> rv(4) = 2
rv =
    3  55  11  2
```

If there is a gap between the end of the vector and the specified element, 0's are filled in. For example, the following extends the variable created earlier again:

```
>> rv(6) = 13
rv =
    3  55  11  2  0  13
```

 PRACTICE 1.6

Think about what would be produced by the following sequence of statements and expressions, and then type them in to verify your answers:

pv = 2:2:8
pv(4) = 33
pv(6) = 11
prac = pv(3:5)
linspace(4,12,3)

1.5.2 Creating Column Vectors

One way to create a column vector is by explicitly putting the values in square brackets, separated by semicolons:

```
>> c = [1; 2; 3; 4]
c =
     1
     2
     3
     4
```

There is no direct way to use the colon operator described earlier to get a column vector. However, any row vector created using any of these methods can be *transposed* to get a column vector. In general, the transpose of a matrix is a new matrix in which the rows and columns are interchanged. For vectors, transposing a row vector results in a column vector, and transposing a column vector results in a row vector. MATLAB has a built-in operator, the apostrophe, to get a transpose.

```
>> r = 1:3;
>> c = r'
c =
     1
     2
     3
```

1.5.3 Creating Matrix Variables

Creating a matrix variable is really just a generalization of creating row and column vector variables. That is, the values within a row are separated by either spaces or commas, and the different rows are separated by semicolons. For example, the matrix variable *mat* is created by explicitly typing values:

```
>> mat = [4 3 1; 2 5 6]
mat =
     4     3     1
     2     5     6
```

There must always be the same number of values in each row. If you attempt to create a matrix in which there are different numbers of values in the rows, the result will be an error message; for example:

```
>> mat = [3 5 7; 1 2]
??? Error using ==> vertcat
CAT arguments dimensions are not consistent.
```

Iterators can also be used for the values on the rows using the colon operator; for example:

```
>> mat = [2:4; 3:5]
mat =
     2     3     4
     3     4     5
```

Different rows in the matrix can also be specified by pressing the Enter key after each row instead of typing a semicolon when entering the matrix values; for example:

```
>> newmat = [2 6 88
33 5 2]

newmat =
     2     6    88
    33     5     2
```

Matrices of random numbers can be created using the **rand** and **randint** functions. The first two arguments to the **randint** function specify the size of the matrix of random integers. For example, the following will create a 2 × 4 matrix of random integers, each in the range from 10 to 30:

```
>> randint(2,4,[10,30])
ans =
    29    22    28    19
    14    20    26    10
```

For the **rand** function, if a single value n is passed to it, an $n \times n$ matrix will be created, or passing two arguments will specify the number of rows and columns:

```
>> rand(2)
ans =
    0.2311    0.4860
    0.6068    0.8913
>> rand(1,3)
ans =
    0.7621  0.4565  0.0185
```

MATLAB also has several functions that create special matrices. For example, the **zeros** function creates a matrix of all zeros. Like **rand**, either one argument can be passed (which will be both the number of rows and columns), or two arguments (first the number of rows and then the number of columns).

```
>> zeros(3)
ans =

        0        0        0
        0        0        0
        0        0        0
>> zeros(2,4)
ans =

        0        0        0        0
        0        0        0        0
```

We will see examples of other special matrix functions in Chapter 11.

1.5.3.1 *Referring to and Modifying Matrix Elements*

To refer to matrix elements, the row and then the column indices are given in parentheses (always the row index first and then the column). For example, this creates a matrix variable *mat*, and then refers to the value in the second row, third column of *mat*:

```
>> mat = [2:4; 3:5]
mat =

        2        3        4
        3        4        5
>> mat(2,3)
ans =
        5
```

It is also possible to refer to a subset of a matrix. For example, this refers to the first and second rows, second and third columns:

```
>> mat(1:2,2:3)
ans =

        3        4
        4        5
```

Using a colon for the row index means all rows, regardless of how many, and using a colon for the column index means all columns. For example, this refers to the entire first row:

```
>> mat(1,:)
ans =
    2  3  4
```

and this refers to the entire second column:

```
>> mat(:, 2)
ans =
        3
        4
```

If a single index is used with a matrix, MATLAB *unwinds* the matrix column by column. For example, for the matrix *intmat* created here, the first two elements are from the first column, and the last two are from the second column:

```
>> intmat = randint(2,2,[0 100])
intmat =
    100         77
     28         14
>> intmat(1)
ans =
    100
>> intmat(2)
ans =
     28
>> intmat(3)
ans =
     77
>> intmat(4)
ans =
     14
```

This is called *linear indexing*. It is usually much better style when working with matrices to refer to the row and column indices, however.

An individual element in a matrix can be modified by assigning a value.

```
>> mat = [2:4; 3:5];
>> mat(1,2) = 11
mat =
     2          11          4
     3           4          5
```

An entire row or column could also be changed. For example, the following replaces the entire second row with values from a vector:

```
>> mat(2,:) = 5:7
mat =
     2          11          4
     5           6          7
```

Notice that since the entire row is being modified, a vector with the correct length must be assigned.

To extend a matrix, an individual element could not be added since that would mean there would no longer be the same number of values in every row. However, an entire row or column could be added. For example, the following would add a fourth column to the matrix:

```
>> mat(:,4) = [9 2]'
mat =
     2         11         4         9
     5          6         7         2
```

Just as we saw with vectors, if there is a gap between the current matrix and the row or column being added, MATLAB will fill in with zeros.

```
>> mat(4,:) = 2:2:8
mat =
     2         11         4         9
     5          6         7         2
     0          0         0         0
     2          4         6         8
```

1.5.4 Dimensions

The **length** and **size** functions in MATLAB are used to find array dimensions. The **length** function returns the number of elements in a vector. The **size** function returns the number of rows and columns in a matrix. For a matrix, the **length** function will return either the number of rows or the number of columns, whichever is largest. For example, the following vector, *vec*, has four elements so its length is 4. It is a row vector, so the size is 1 × 4.

```
>> vec = -2:1
vec =
   -2  -1   0   1
>> length(vec)
ans =
    4
>> size(vec)
ans =
    1   4
```

For the matrix *mat* shown next, it has three rows and two columns, so the size is 3 × 2. The length is the larger dimension, 3.

```
>> mat = [1:3; 5:7]'
mat =
    1          5
    2          6
    3          7
>> size(mat)
ans =
    3   2
```

```
>> length(mat)
ans =
   3
>> [r c] = size(mat)
r =
   3
c =
   2
```

Note: The last example demonstrates a very important and unique concept in MATLAB: the ability to have a vector of variables on the left-hand side of an assignment.

The **size** function returns two values, so in order to capture these values in separate variables we put a vector of two variables on the left of the assignment. The variable *r* stores the first value returned, which is the number of rows, and *c* stores the number of columns.

QUICK QUESTION!

How could you create a matrix of zeros with the same size as another matrix?

Answer:

For a matrix variable *mat*, the following expression would accomplish this:

```
zeros(size(mat))
```

The **size** function returns the size of the matrix, which is then passed to the **zeros** function, which then returns a matrix of zeros with the same size as *mat*. It is not necessary in this case to store the values returned from the **size** function.

MATLAB also has a function, **numel**, which returns the total number of elements in any array (vector or matrix):

```
>> vec = 9:-2:1
vec =
     9    7    5    3    1
>> numel(vec)
ans =
     5
>> mat = randint(2,3,[1,10])
mat =
     7         9         8
     4         6         5
>> numel(mat)
ans =
     6
```

For vectors, this is equivalent to the length of the vector. For matrices, it is the product of the number of rows and columns.

MATLAB also has a built-in expression **end** that can be used to refer to the last element in a vector; for example, v(end) is equivalent to v(length(v)). For matrices, it can refer to the last row or column. So, using **end** for the row index would refer to the last row. In this case, the element referred to is in the first column of the last row:

```
>> mat = [1:3; 4:6]'
mat =
        1        4
        2        5
        3        6
>> mat(end,1)
ans =
        3
```

Using **end** for the column index would refer to the last column (e.g., the last column of the second row):

```
>> mat(2,end)
ans =
        5
```

This can be used only as an index.

1.5.4.1 Changing Dimensions

In addition to the transpose operator, MATLAB has several built-in functions that change the dimensions or configuration of matrices, including **reshape**, **fliplr**, **flipud**, and **rot90**.

The **reshape** function changes the dimensions of a matrix. The following matrix variable *mat* is 3 × 4, or in other words it has 12 elements.

```
>> mat = randint(3,4,[1,100])
mat =
        14        61         2        94
        21        28        75        47
        20        20        45        42
```

These 12 values instead could be arranged as a 2 × 6 matrix, 6 × 2, 4 × 3, 1 × 12, or 12 × 1. The **reshape** function iterates through the matrix columnwise. For example, when reshaping *mat* into a 2 × 6 matrix, the values from the first column in the original matrix (14, 21, and 20) are used first, then the values from the second column (61, 28, 20), and so forth.

```
>> reshape(mat,2,6)
ans =
```

14	20	28	2	45	47
21	61	20	75	94	42

The **fliplr** function "flips" the matrix from left to right (in other words the left-most column, the first column, becomes the last column and so forth), and the **flipud** functions flips up to down. Note that in these examples *mat* is unchanged; instead, the results are stored in the default variable *ans* each time.

```
>> mat = randint(3,4,[1 100])
mat =
```

14	61	2	94
21	28	75	47
20	20	45	42

```
>> fliplr(mat)
ans =
```

94	2	61	14
47	75	28	21
42	45	20	20

```
>> mat
mat =
```

14	61	2	94
21	28	75	47
20	20	45	42

```
>> flipud(mat)
ans =
```

20	20	45	42
21	28	75	47
14	61	2	94

The **rot90** function rotates the matrix counterclockwise 90 degrees, so for example the value in the top-right corner becomes instead the top-left corner and the last column becomes the first row:

```
>> mat
mat =
```

14	61	2	94
21	28	75	47
20	20	45	42

```
>> rot90(mat)
```

```
ans =
        94      47      42
         2      75      45
        61      28      20
        14      21      20
```

QUICK QUESTION!

Is there a **rot180** function? Is there a **rot-90** function (to rotate clockwise)?

Answer:

Not exactly, but a second argument can be passed to the **rot90** function, which is an integer *n*; the function will rotate 90*n* degrees. The integer can be positive or negative. For example, if 2 is passed, the function will rotate the matrix 180 degrees (so, it would be the same as rotating the value of *ans* another 90 degrees).

```
>> mat
mat =
        14      61       2      94
        21      28      75      47
        20      20      45      42
>> rot90(mat,2)
ans =
```

```
        42      45      20      20
        47      75      28      21
        94       2      61      14
```

If a negative number is passed for *n*, the rotation would be in the opposite direction, in other words, clockwise.

```
>> mat
mat =
        14      61       2      94
        21      28      75      47
        20      20      45      42
>> rot90(mat,-1)
ans =
        20      21      14
        20      28      61
        45      75       2
        42      47      94
```

The function **repmat** can also be used to create a matrix; **repmat(mat,m,n)** creates a larger matrix, which consists of an *m* × *n* matrix of copies of mat. For example, here is a 2 × 2 random matrix:

```
>> intmat = randint(2,2,[0 100])
intmat =
       100      77
        28      14
```

The function **repmat** can be used to replicate this matrix six times as a 3 × 2 matrix of the variable *intmat*.

```
>> repmat(intmat,3,2)
ans =
       100      77     100      77
        28      14      28      14
       100      77     100      77
```

28	14	28	14
100	77	100	77
28	14	28	14

1.5.5 Using Functions with Vectors and Matrices

Since MATLAB is written to work with vectors and matrices, an entire vector or matrix can be passed as an argument to a function. MATLAB will evaluate the function on every element, and return as a result a vector or matrix with the same dimensions as the original. For example, we could pass the following vector, *vec*, to the **abs** function in order to get the absolute value of every element.

```
>> vec = -3:4
vec =
    -3    -2    -1     0     1     2     3     4
>> abs(vec)
ans =
     3     2     1     0     1     2     3     4
```

The original vector *vec* has eight elements, and since the **abs** function is evaluated for every element, the resulting vector also has eight elements.

This also would be the case for matrices:

```
>> mat = randint(2,3,[-5,5])
mat =
    -5    -1     0
     3     5    -1
>> abs(mat)
ans =
     5     1     0
     3     5     1
```

We will see much more on operations and functions of arrays (vectors and matrices) in Chapters 4 and 11.

1.5.6 Empty Vectors

An *empty vector*, or, in other words, a vector that stores no values, can be created using empty square brackets:

```
>> evec = []
evec =
     []
>> length(evec)
ans =
     0
```

Then, values can be added to the vector by *concatenating*, or adding values to the existing vector. The following statement takes what is currently in *evec*, which is nothing, and adds a 4 to it.

```
>> evec = [evec 4]
evec =
     4
```

The following statement takes what is currently in *evec*, which is 4, and adds an 11 to it.

```
>> evec = [evec 11]
evec =
     4  11
```

This can be continued as many times as desired, in order to build a vector up from nothing.

Empty vectors can also be used to *delete elements from arrays*. For example, to remove the third element from an array, the empty vector is assigned to it:

```
>> vec = 1:5
vec =
     1     2     3     4     5
>> vec(3) = []
vec =
     1     2     4     5
```

The elements in this vector are now numbered 1 through 4.

Subsets of a vector could also be removed; for example:

```
>> vec = 1:8
vec =
     1     2     3     4     5     6     7     8
>> vec(2:4) = []
vec =
     1     5     6     7     8
```

Individual elements cannot be removed from matrices, since matrices always have to have the same number of elements in every row.

```
>> mat = [7 9 8; 4 6 5]
mat =
     7     9     8
     4     6     5
>> mat(1,2) = [];
??? Indexed empty matrix assignment is not allowed.
```

However, entire rows or columns could be removed from a matrix. For example, to remove the second column:

```
>> mat(:,2) = []
mat =

     7          8
     4          5
```

SUMMARY

Common Pitfalls

It is common when learning to program to make simple spelling mistakes and to confuse the necessary punctuation. Following are examples of very common errors:

- Putting a space in a variable name
- Confusing the format of an assignment statement as

```
expression = variablename
```

rather than

```
variablename = expression
```

The variable name must always be on the left.

- Using a built-in function name as a variable name, and then trying to use the function
- Confusing the two division operators / and \
- Forgetting the operator precedence rules
- Confusing the order of arguments passed to functions, for example, to find the remainder of dividing 3 into 10 using **rem(3,10)** instead of **rem(10,3)**
- Not using different types of arguments when testing functions
- Attempting to create a matrix that does not have the same number of values on each row
- Forgetting to use parentheses to pass an argument to a function; for example, **fix 2.3** instead of **fix(2.3)**. MATLAB returns the ASCII equivalent for each character when this mistake is made. (What happens is that is that it is interpreted as the function of a string; for example, **fix('2.3')**).

Programming Style Guidelines

Following these guidelines will make your code much easier to read and understand, and therefore easier to work with and modify.

- Use mnemonic variable names (names that make sense; for example, *radius* instead of *xyz*).

 PRACTICE 1.7

Think about what would be produced by the following sequence of statements and expressions, and then type them to verify your answers.

```
m = [1:4; 3 11 7 2]
m(2,3)
m(:,3)
m(4)
size(m)
numel(m)
reshape(m,1,numel(m))
vec = m(1,:)
vec(2) = 5
vec(3) = []
vec(5) = 8
vec = [vec 11]
```

- Do not use names of built-in functions as variable names.
- If different sets of random numbers are desired, set the seed for the **rand** function.
- Do not use just a single index when referring to elements in a matrix; instead, use both the row and column indices.
- To be general, never assume that the dimensions of any array (vector or matrix) are known. Instead, use the function **length** to determine the number of elements in a vector, and the function **size** for a matrix, for example:

```
len = length(vec);
[r c] = size(mat);
```

MATLAB Functions and Commands

info	floor	double	linspace
demo	ceil	int8	zeros
help	round	int16	length
lookfor	rem	int32	size
namelengthmax	sign	int64	numel
who	pi	intmin	end
whos	i	intmax	reshape
clear	j	char	fliplr
format	inf	logical	flipud
sin	exp	rand	rot90
abs	NaN	clock	repmat
fix	single	randint	

MATLAB Operators

assignment =	multiplication *	divided into \	colon:
addition +	exponentiation ^	parentheses ()	transpose '
subtraction −	divided by /	negation -	

Exercises

1. Create a variable, *myage*, and store your age in it. Subtract one from the value of the variable. Add two to the value of the variable.
2. Use the built-in function **namelengthmax** to find out the maximum number of characters that you can have in an identifier name under your version of MATLAB.
3. Explore the **format** command in more detail. Use **help format** to find options. Experiment with **format bank** to display dollar values.
4. Find a **format** option that would result in the following output format:

```
>> 5/16 + 2/7
ans =
     67/112
```

5. Think about what the results would be for the following expressions, and then type them to verify your answers.
```
25 / 4 * 4
```

```
3 + 4 ^ 2
4 \ 12 + 4
   3 ^ 2
(5 - 2) * 3
```

6. Create a variable, *pounds*, to store a weight in pounds. Convert this to kilograms and assign the result to a variable *kilos*. The conversion factor is 1 kilogram = 2.2 pounds.

7. The combined resistance RT of three resistors R_1, R_2, and R_3 in parallel is given by

$$RT = \frac{1}{\dfrac{1}{R_1} + \dfrac{1}{R_2} + \dfrac{1}{R_3}}$$

Create variables for the three resistors and store values in each, and then calculate the combined resistance.

8. Create a variable *ftemp* to store a temperature in degrees Fahrenheit (F). Convert this to degrees Celsius and store the result in a variable *ctemp*. The conversion factor is C = (F − 32) * 5/9.

9. The function **sin** calculates and returns the sine of an angle in radians. Use **help elfun** to find the name of the function that returns the sine of an angle in degrees. Verify that calling this function and passing 90 degrees to it results in 1.

10. A vector can be represented by its rectangular coordinates x and y or by its polar coordinates r and θ. The relationship between them is given by the equations:

```
x = r * cos(θ)
y = r * sin(θ)
```

Assign values for the polar coordinates to variables *r* and *theta*. Then, using these values, assign the corresponding rectangular coordinates to variables *x* and *y*.

11. Wind often makes the air feel even colder than it is. The Wind Chill Factor (WCF) measures how cold it feels with a given air temperature T (in degrees Fahrenheit) and wind speed (V, in miles per hour). One formula for the WCF is:

```
WCF=35.7+0.6T-35.7(V0.16)+0.43T(V0.16)
```

Create variables for the temperature T and wind speed V, and then using this formula calculate the WCF.

12. Use **help elfun** or experiment to answer the following questions:
   - Is **fix(3.5)** the same as **floor(3.5)**?
   - Is **fix(3.4)** the same as **fix(-3.4)**?
   - Is **fix(3.2)** the same as **floor(3.2)**?
   - Is **fix(-3.2)** the same as **floor(-3.2)**?
   - Is **fix(-3.2)** the same as **ceil(-3.2)**?

13. Find MATLAB expressions for the following:

$\sqrt{19}$

$3^{1.2}$

$\tan(\pi)$

14. Use **intmin** and **intmax** to determine the range of values that can be stored in the types **int32** and **int64**.

15. Are there equivalents to **intmin** and **intmax** for real number types? Use **help** to find out.

16. Store a number with a decimal place in a **double** variable (the default). Convert the value of that variable to the type **int32** and store the result in a new variable.

17. Generate a random:

- Real number in the range from 0 to 1
- Real number in the range from 0 to 20
- Real number in the range from 20 to 50
- Integer in the range from 0 to 10
- Integer in the range from 0 to 11
- Integer in the range from 50 to 100

18. Get into a new Command Window, and type **rand** to get a random real number. Make a note of the number. Then, exit MATLAB and repeat this, again making note of the random number; it should be the same as before. Finally, exit MATLAB and again get into a new Command Window. This time, change the seed before generating a random number; it should be different.

19. In the ASCII character encoding, the letters of the alphabet are in order: 'a' comes before 'b' and also 'A' comes before 'B', for example. However, which comes first: lower- or uppercase letters?

20. Shift the string 'xyz' up in the character encoding by two characters.

21. Using the colon operator, create the following vectors

3	4	5	6	
1.0000	1.5000	2.0000	2.5000	3.0000
5	4	3	2	

22. Using the **linspace** function, create the following vectors:

4	6	8		
-3	-6	-9	-12	-15
9	7	5		

23. Create the following vectors twice, using **linspace** and using the colon operator:

1	2	3	4	5	6	7	8	9	10
2	7	12							

24. Create a variable, *myend*, which stores a random integer in the range from 8 to 12. Using the colon operator, create a vector that iterates from 1 to *myend* in steps of 3.

25. Using the colon operator and the transpose operator, create a column vector that has the values −1 to 1 in steps of 0.2.

26. Write an expression that refers to only the odd numbered elements in a vector, regardless of the length of the vector. Test your expression on vectors that have both an odd and even number of elements.

27. Create a vector variable, *vec*; it can have any length. Then, write assignment statements that would store the first half of the vector in one variable and the second half in another. Make sure that your assignment statements are general, and work whether *vec* has an even or odd number of elements (hint: use a rounding function such as **fix**).

28. Using colon operators for the rows, create the matrix:

7	6	5
3	5	7

29. Generate a 2 × 3 matrix of random

   ■ Real numbers, each in the range from 0 to 1
   ■ Real numbers, each in the range from 0 to 10
   ■ Integers, each in the range from 5 to 20

30. Create a variable, *rows*, which is a random integer in the range from 1 to 5. Create a variable, *cols*, which is a random integer in the range from 1 to 5. Create a matrix of all zeros with the dimensions given by the values of *rows* and *cols*.

31. Find an efficient way to generate the following matrix:
    mat =

7	8	9	10
12	10	8	6

   ■ Then, give expressions that will, for the matrix *mat*,
     ■ Refer to the element in the first row, third column
     ■ Refer to the entire second row
     ■ Refer to the first two columns

32. Create a matrix variable, *mymat*, which stores the following:
    mymat =

2	5	8
7	5	3

Using this matrix, find a simple expression that will transform the matrix into each of the following:

2	7
5	5
8	3

```
2     5
7     8
5     3


8     5     2
3     5     7


8     3
5     5
2     7


2     5     8     2     5     8
7     5     3     7     5     3
```

33. Create a 4 × 2 matrix of all zeros and store it in a variable. Then, replace the second row in the matrix with a 3 and a 6.
34. Create a vector, x, which consists of 20 equally spaced points in the range from −π to +π. Create a y vector that is **sin(x)**.
35. Create a 3 × 5 matrix of random integers, each in the range from −5 to 5. Get the **sign** of every element.
36. Create a 4 × 6 matrix of random integers, each in the range from −5 to 5; store it in a variable. Create another matrix that stores for each element the absolute value of the corresponding element in the original matrix.
37. Create a 3 × 5 matrix of random real numbers. Delete the third row.
38. The built-in function **clock** returns a vector that contains six elements: the first three are the current date (year, month, day) and the last three represent the current time in hours, minutes, and seconds. The seconds is a real number, but all others are integers. Store the result from **clock** in a variable called *myc*. Then, store the first three elements from this variable in a variable called *today*, and the last three elements in a variable called *now*. Use the **fix** function on the vector variable called *now* to get just the integer part of the current time.

CHAPTER 2

Introduction to MATLAB Programming

KEY WORDS

computer program
script
algorithm
top-down design
external file
user
default input device
prompting
default output device
execute/run
high-level language
machine language
executable
compiler
source code

object code
interpreter
script file
documentation
comment
input/output (I/O)
empty string
error message
formatting
format string
placeholder
conversion character
field width
leading blank
trailing blank

exponential notation
plot symbols
markers
line type
toggle
mode
appending
control
user-defined function
function definition
function header
function body
output argument
input argument
local variable

CONTENTS

We have now used the MATLAB® product interactively in the Command Window. That is sufficient when all that is needed is a simple calculation. However, in many cases, many steps are required before the final result can be obtained. In those cases, it is more convenient to group statements together in what is called a *computer program*.

In this chapter, we will introduce the simplest MATLAB® programs, called *scripts*. Examples of scripts that customize simple plots will illustrate the concept. Input will be introduced, both from files and from the user. Output, to files and to the screen, will also be introduced. Finally, user-defined functions that calculate and return values will be described. These topics serve as an introduction to programming, which will be explored further in Chapter 5.

2.1 ALGORITHMS

Before writing any computer program, it is useful to first outline the steps that will be necessary. An *algorithm* is the sequence of steps needed to solve a problem. In a modular approach to programming, the problem solution is broken down into separate steps, and then each step is further refined until the resulting steps are small enough to be manageable tasks. This is called the *top-down design* approach.

As a simple example, consider the problem of calculating the area of a circle. First, it is necessary to determine what information is needed in order to solve the problem, which in this case is the radius of the circle. Next, given the radius of the circle, the area of the circle would be calculated. Last, once the area has been calculated, it has to be displayed in some way. So, the basic algorithm is:

- Get the input: the radius
- Calculate the result: the area
- Display the output

Even with an algorithm this simple, it is possible to refine each of the steps further. When a program is written to implement this algorithm, the steps would be:

1. Where does the input come from? Two possible choices would be from an *external file* on a disk, or from the *user* (the person who is running the program), who enters the number by typing it from the keyboard. For every system, one of these will be the *default input device* (which means, if not specified otherwise, this is where the input comes from). If the user is supposed to enter the radius, the user has to be told to type it (and, in what units). Telling the user what to enter is called *prompting*. So, the input step actually becomes two steps: prompt the user to enter a radius, and then read it into the program.

2. To calculate the area, the formula is needed. In this case, the area of the circle is π multiplied by the radius squared. So, that means the value of π is needed in the program, which is an example of a constant.

3. Where does the output go? Two possibilities are (1) to an external file, or (2) to a window on the screen. Depending on the system, one of these will be the *default output device*. When displaying the output from the program, it should always be as informative as possible. In other words, instead of just printing the area (just the number), it should be printed in a nice sentence format. Also, to make the output even more clear, the input should be printed. For example, the output might be the sentence: "For a circle with a radius of 1 inch, the area is 3.1416 inches squared."

For most programs, the basic algorithm consists of the three steps just outlined:

1. Get the input(s).

2. Calculate the result(s).

3. Display the result(s).

As can be seen here, even the simplest problem solutions can then be further refined.

2.2 MATLAB SCRIPTS

Once a problem has been analyzed, and the algorithm for its solution has been written and refined, the algorithm then is translated into a particular programming language. A *computer program* is a sequence of instructions, in a given language, that accomplishes a task. To *execute*, or *run*, a program is to have the computer actually follow these instructions.

High-level languages have English-like commands and functions, such as "print this" or "if x < 5 do something." The computer, however, can interpret commands only written in its *machine language*. Programs that are written in high-level languages must therefore be translated into machine language before the computer can actually execute the sequence of instructions in the program. A program that does this translation from a high-level language to an *executable* file is called a *compiler*. The original program is called the *source code*, and the resulting executable program is called the *object code*.

By contrast, an *interpreter* goes through the code line-by-line, executing each command as it goes. MATLAB uses either what are called *script* files or M-files (the reason for this is that the extension on the filename is .m). These script files are interpreted, rather than compiled. Therefore, the correct terminology is that these are scripts, and not programs. However, the terms are somewhat loosely used by many people, and the documentation in MATLAB itself refers to scripts as programs. In this book, we will reserve the use of the word "program" to mean a set of scripts and functions, as described briefly later in this chapter and then in more detail in Chapter 5.

A script is a sequence of MATLAB instructions that is stored in a file and saved. The contents of a script can be displayed in the Command Window using the **type** command. The script can be executed, or run, by simply entering the name of the file (without the .m extension).

To create a script, click File, then New, then M-file. A new window will appear called the Editor. To create a new script, simply type the sequence of statements (notice that line numbers will appear on the left).

When finished, save the file using File and then Save. Make sure that the extension .m is on the filename (this should be the default). The rules for filenames are the same as for variables (they must start with a letter, after that there can be letters, digits, or the underscore, etc.). By default, scripts will be saved in the Work Directory. If you want to save the file in a different directory, the Current Directory can be changed.

For example, we will now create a script called script1.m that calculates the area of a circle. It assigns a value for the radius, and then calculates the area based on that radius.

In this text, scripts will be displayed in a box with the name on top.

script1.m

```
radius = 5
area = pi * (radius^2)
```

In the Command Window, the contents of the script can be displayed, and the script can be executed. The **type** command shows the contents of the file named script1.m (notice that the .m is not included):

```
>> type script1
radius = 5
area = pi * (radius^2)
```

There are two ways to view a script once it has been written: either open the Editor Window to view it, or use the **type** command as shown here to display it in the Command Window.

To actually run or execute the script, the name of the file is entered at the prompt (again, without the .m). When executed, the results of the two assignment statements are displayed, since the output was not suppressed for either statement.

```
>> script1
radius =
     5
area =
    78.5398
```

Once the script has been executed, you may find that you want to make changes to it (especially if there are errors!). To edit an existing file, there are several methods to open it. The easiest are:

- Click File, then Open, then click the name of the file.
- Click the Current Directory tab (if it is not already shown), then double-click the name of the file.

2.2.1 Documentation

It is very important that all scripts be *documented* well, so that people can understand what the script does and how it accomplishes that. One way of documenting a script is to put *comments* in it. In MATLAB, a comment is anything from a % to the end of that particular line. Comments are completely ignored when the script is executed. To put in a comment, simply type the % symbol at the beginning of a line, or select the comment lines and then click Text and then Comment and the Editor will put in the % symbols for you.

For example, the previous script to calculate the area of a circle could be modified to have comments:

```
script1b.m
% This program calculates the area of a circle

% First the radius is assigned
radius = 5
% The area is calculated based on the radius
area = pi * (radius^2)
```

The first comment at the beginning of the script describes what the script does. Then, throughout the script, comments describe different parts of the script (not usually a comment for every line, however!). Comments don't affect what a script does, so the output from this script would be the same as for the previous version.

The **help** command in MATLAB works with scripts as well as with built-in functions. The first block of comments (defined as contiguous lines at the beginning) will be displayed. For example, for script1b:

```
>> help script1b
   This program calculates the area of a circle
```

The reason that a blank line was inserted in the script between the first two comments is that otherwise both would have been interpreted as one contiguous comment, and both lines would have been displayed with **help**. The very first comment line is called the H1 line; it is what the function **lookfor** searches through.

PRACTICE 2.1

Write a script to calculate the area of a rectangle. Be sure to comment the script.

2.3 INPUT AND OUTPUT

The previous script would be much more useful if it were more general; for example, if the value of the radius could be read from an external source rather than being assigned in the script. Also, it would be better to have the script print the output in a nice, informative way. Statements that accomplish these tasks are called *input/output* statements, or *I/O* for short. Although for simplicity examples of input and output statements will be shown here from the Command Window, these statements will make the most sense in scripts.

2.3.1 Input Function

Input statements read in values from the default or *standard input device*. In most systems, the default input device is the keyboard, so the input statement reads in values that have been entered by the *user*, or the person who is running the script. In order to let the user know what he or she is supposed to enter, the script must first *prompt* the user for the specified values.

The simplest input function in MATLAB is called **input**. The **input** function is used in an assignment statement. To call it, a string is passed, which is the prompt that will appear on the screen, and whatever the user types will be stored in the variable named on the left of the assignment statement. To make it easier to read the prompt, put a colon and then a space after the prompt. For example,

```
>> rad = input('Enter the radius: ')
Enter the radius: 5
rad =
    5
```

If character or string input is desired, 's' must be added after the prompt:

```
>> letter = input('Enter a char: ','s')
Enter a char: g
letter =
g
```

Notice that although this is a string variable, the quotes are not shown. However, they are shown in the Workspace Window.

If the user enters only spaces or tabs before pressing the Enter key, they are ignored and an *empty string* is stored in the variable:

```
>> mychar = input('Enter a character: ', 's')
Enter a character:
mychar =
    ' '
```

Notice that in this case the quotes are shown, to demonstrate that there is nothing inside of the string.

However, if blank spaces are entered before other characters, they are included in the string. In this example, the user pressed the space bar four times before entering "go":

```
>> mystr = input('Enter a string: ', 's')
Enter a string:      go
mystr =
     go
>> length(mystr)
ans =
       6
```

QUICK QUESTION!

What would be the result if the user enters blank spaces after other characters? For example, the user here entered "xyz   " (four blank spaces):

```
>> mychar = input('Enter chars:', 's')
Enter chars: xyz
mychar =
xyz
```

Answer:
The space characters would be stored in the string variable. It is difficult to see in this example, but is clear from the length of the string.

```
>> length(mychar)
ans =
       7
```

The length can be seen in the Command Window by using the mouse to highlight the value of the variable; the xyz and four spaces will be highlighted.

It is also possible for the user to type quotation marks around the string rather than including the second argument 's' in the call to the **input** function:

```
>> name = input('Enter your name: ');
Enter your name: 'Stormy'
```

However, it is better to signify that character input is desired in the **input** function itself.

Normally, the results from **input** statements are suppressed with a semicolon at the end of the assignment statements, as shown here. Notice what happens if string input has not been specified, but the user enters a letter rather than a number:

```
>> num = input('Enter a number: ')
Enter a number: t
??? Error using ==> input
```

```
Undefined function or variable 't'.
Enter a number: 3
num =
     3
```

MATLAB gave an *error message* and repeated the prompt. However, if *t* is the name of a variable, MATLAB will take its value as the input:

```
>> t = 11;
>> num = input('Enter a number: ')
Enter a number: t
num =
    11
```

Separate **input** statements are necessary if more than one input is desired. For example

```
>> x = input('Enter the x coordinate: ');
>> y = input('Enter the y coordinate: ');
```

 PRACTICE 2.2

Create a script that would prompt the user for a temperature, and then 'F' or 'C', and store both inputs in variables. For example, when executed it would look like this (assuming the user enters 85 and then F):

```
Enter the temperature: 85
Is that F or C?: F
```

2.3.2 Output Statements: disp and fprintf

Output statements display strings and the results of expressions, and can allow for *formatting*, or customizing how they are displayed. The simplest output function in MATLAB is **disp**, which is used to display the result of an expression or a string without assigning any value to the default variable *ans*. However, **disp** does not allow formatting. For example,

```
>> disp('Hello')
Hello
>> disp(4^3)
    64
```

Formatted output can be printed to the screen using the **fprintf** function. For example,

```
>> fprintf('The value is %d, for sure!\n',4^3)
The value is 64, for sure!
```

To the **fprintf** function, first a string (called the *format string*) is passed, which contains any text to be printed as well as formatting information for the expressions to be printed. In this example, the %d is an example of format information. The %d is sometimes called a *placeholder*; it specifies where the value of the expression that is after the string is to be printed. The character in the placeholder is called the *conversion character*, and it specifies the type of value that is being printed. There are others, but what follows is a list of the simple placeholders:

```
%d integers (it actually stands for decimal integer)
%f floats
%c single characters
%s strings
```

Don't confuse the % in the placeholder with the symbol used to designate a comment.

The character '\n' at the end of the string is a special character called the *newline* character; when it is printed the output moves down to the next line.

QUICK QUESTION!

What do you think would happen if the newline character were omitted from the earlier **fprintf** statement?

Answer:

Without it, the next prompt would end up on the same line as the output. It is still a prompt, and so an expression can be entered, but it looks messy, as shown here:

```
>> fprintf('The value is %d, surely!',...
4^3)
The value is 64, surely!>> 5 + 3
```

```
ans =
      8
```

Note that with the **disp** function, however, the prompt will always appear on the next line:

```
>> disp('Hi')
Hi
>>
```

QUICK QUESTION!

How can you get a blank line in the output?

Answer:

Have two newline characters in a row.

```
>> fprintf('The value is %d, \n\nOK!\n', 4^3)
The value is 64,

OK!
```

This also points out that the newline character can be anywhere in the string; when it is printed, the output moves down to the next line.

Note: The newline character can also be used in the prompt in the **input** statement; for example:

```
>> x = input('Enter the \nx coordinate: ');
Enter the
x coordinate: 4
```

However, that is the only formatting character allowed in the prompt in **input**.

A *field width* can also be included in the placeholder in **fprintf**, which specifies how many characters total are to be used in printing. For example, %5d would indicate a field width of 5 for printing an integer and %10s would indicate a field width of 10 for a string. For floats, the number of decimal places can also be specified; for example, %6.2f means a field width of 6 (including the decimal point and the decimal places) with two decimal places. For floats, just the number of decimal places can also be specified; for example, %.3f indicates three decimal places.

```
>> fprintf('The int is %3d and the float is %6.2f\n',5,4.9)
The int is   5 and the float is   4.90
```

Note that if the field width is wider than necessary, *leading* blanks are printed, and if more decimal places are specified than necessary, *trailing* zeros are printed.

QUICK QUESTION!

What do you think would happen if you tried to print 1234.5678 in a field width of 3 with two decimal places, for example using

```
>> fprintf('%3.2f\n', 1234.5678)
```

Answer:
It would print the entire 1234, but round the decimals to two places, for example:

```
1234.57
```

If the field width is not large enough to print the number, the field width will be increased. Basically, to cut the number off would give a misleading result, but rounding the decimal places does not change the number by much.

QUICK QUESTION!

What would happen if you use the %d conversion character but you're trying to print a real number?
Answer:
MATLAB will show the result using *exponential notation*:

```
>> fprintf('%d\n',1234567.89)
1.234568e+006
```

Note: If you want exponential notation, this is not the correct way to get it; instead, there are conversion characters that can be used. Use the **help** browser to see this option as well as many others!

There are many other options for the format string. For example, the value being printed can be left-justified within the field width using a minus sign. The following example shows the difference between printing the integer 3 using %5d and using %–5d. The x's are just used to show the spacing.

```
>> fprintf('The integer is xx%5dxx and xx%-5dxx\n',3,3)
The integer is xx    3xx and xx3    xx
```

Also, strings can be truncated by specifying decimal places:

```
>> fprintf('The string is %s or %.4s\n', 'truncate',...
'truncate')
The string is truncate or trun
```

There are several special characters that can be printed in the format string in addition to the newline character. To print a slash, two slashes in a row are used, and also to print a single quote two single quotes in a row are used. Additionally, \t is the tab character.

```
>> fprintf('Try this out: tab\t quote '' slash \\ \n')
Try this out: tab    quote ' slash \
```

2.3.2.1 *Printing Vectors and Matrices*

For a vector, if a conversion character and the newline character are in the format string, it will print in a column regardless of whether the vector itself is a row vector or a column vector.

```
>> vec = 2:5;
>> fprintf('%d\n', vec)
2
3
4
5
```

Note that this was a row vector. A column vector would print exactly the same way, as a column. Without the newline character, it would print in a row but the next prompt would appear on the same line:

```
>> fprintf('%d', vec)
2345>>
```

However, in a script, a separate newline character could be printed to avoid this problem.

printvec.m

```
% This demonstrates printing a vector
vec = 2:5;
fprintf('%d',vec)
fprintf('\n')
```

```
>> printvec
2345
>>
```

If the number of elements in the vector is known, that many conversion characters can be specified and then the newline:

```
>> fprintf('%d %d %d %d\n', vec)
2 3 4 5
```

This is not very general, however, and is therefore not preferable.

For matrices, MATLAB unwinds the matrix column by column. For example, consider the following random 2 × 3 matrix:

```
>> mat = randint(2,3,[1,10])
mat =
     5   9   8
     4   1   10
```

Specifying one conversion character and then the newline character will print the elements from the matrix in one column. The first values printed are from the first column, then the second column, and so on.

```
>> fprintf('%d\n', mat)
5
4
9
1
8
10
```

If three of the %d conversion characters are specified, the **fprintf** will print three numbers across on each line of output, but again the matrix is unwound column by column. It again prints first the two numbers from the first column (across on the first row of output), then the first value from the second column, and so on.

```
>> fprintf('%d %d %d\n', mat)
5   4   9
1   8   10
```

If the transpose of the matrix is printed, however, using the three %d conversion characters, the matrix is printed as it appears when created.

```
>> fprintf('%d %d %d\n', mat') % Note the transpose
5   9   8
4   1   10
```

For vectors and matrices, even though formatting cannot be specified, the **disp** function may be easier to use in general than **fprintf** because it displays the result in a straightforward manner. For example,

```
>> mat = randint(2,3,[5 15])
mat =
      15      11      14
       7      10      13

>> disp(mat)
      15      11      14
       7      10      13
>> vec = 2:5
vec =
       2       3       4       5

>> disp(vec)
       2       3       4       5
```

2.4 SCRIPTS WITH INPUT AND OUTPUT

Putting all this together, we can implement the algorithm from the beginning of this chapter. The following script calculates and prints the area of a circle. It first prompts the user for a radius, reads in the radius, and then calculates and prints the area of the circle based on this radius.

script2.m

```
% This script calculates the area of a circle
% It prompts the user for the radius

% Prompt the user for the radius and calculate
% the area based on that radius
radius = input('Please enter the radius: ');
area = pi * (radius^2);
% Print all variables in a sentence format
fprintf('For a circle with a radius of %.2f,',radius)
fprintf('the area is %.2f\n',area)
```

Executing the script produces the following output:

```
>> script2
Please enter the radius: 3.9
For a circle with a radius of 3.90, the area is 47.78
```

Notice that the output from the first two assignment statements is suppressed by putting semicolons at the end. That is frequently done in scripts, so that the exact format of what is displayed by the program is controlled by the **fprintf** functions.

PRACTICE 2.3

Write a script to prompt the user for a character and a number, and print the character in a field-width of 4 and the number left-justified in a field width of 5 with two decimal places. Test this by entering numbers with varying widths.

2.5 SCRIPTS TO PRODUCE AND CUSTOMIZE SIMPLE PLOTS

MATLAB has many graphing capabilities. In many cases, customizing plots is desired and this is easiest to accomplish by creating a script rather than typing one command at a time in the Command Window. For that reason, simple plots and how to customize them will be introduced in this chapter on MATLAB programming.

Typing **help** will display the help topics that contain graph functions, including **graph2d** and **graph3d**. Typing **help graph2d** would display some of the two-dimensional graph functions, as well as functions to manipulate the axes and to put labels and titles on the graphs.

2.5.1 The Plot Function

For now, we'll start with a very simple graph of one point using the **plot** function.

The following script, *plotonepoint*, plots one point. To do this, first values are given for the x and y coordinates of the point in separate variables. The point is then plotted using a red*. The plot is then customized by specifying the minimum and maximum values on first the x- and then y-axis. Labels are then put on the x-axis, the y-axis, and the graph itself using the function **xlabel**, **ylabel**, and **title**. All this can be done from the Command Window, but it is much easier to use a script. The following shows the contents of the script *plotonepoint* that accomplishes this. The x-coordinate represents the time of day (e.g., 11am) and the y-coordinate represents the temperature in degrees Fahrenheit at that time:

plotonepoint.m

```
% This is a really simple plot of just one point!

% Create coordinate variables and plot a red '*'
x = 11;
y = 48;
plot(x,y,'r*')
% Change the axes and label them
axis([9 12 35 55])
xlabel('Time')
ylabel('Temperature')
% Put a title on the plot
title('Time and Temp')
```

In the call to the **axis** function, one vector is passed. The first two values are the minimum and maximum for the x-axis, and the last two are the minimum and maximum for the y-axis. Executing this script brings up a Figure Window with the plot (see Figure 2.1).

To be more general, the script could prompt the user for the time and temperature, rather than just assigning values. Then, the axis function could be used based on whatever the values of x and y are, for example,

```
axis([x-2 x+2 y-10 y+10])
```

In order to plot more than one point, x and y vectors are created to store the values of the (x,y) points. For example, to plot the points

```
(1,1)
(2,5)
(3,3)
(4,9)
(5,11)
(6,8)
```

first an *x* vector is created that has the x values (since they range from 1 to 6 in steps of 1, the colon operator can be used) and then a *y* vector is created with the y values. This will create (in the Command Window) x and y vectors and then plot them (see Figure 2.2).

```
>> x = 1:6;
>> y = [1 5 3 9 11 8];
>> plot(x,y)
```

Notice that the points are plotted with straight lines drawn in between. Also, the axes are set up according to the data; for example, the x values range from 1 to 6 and the y values from 1 to 11, so that is how the axes are set up.

Also, notice that in this case the x values are the indices of the y vector (the y vector has six values in it, so the indices iterate from 1 to 6). When this is the case, it is not necessary to create the x vector. For example,

```
>> plot(y)
```

will plot exactly the same figure without using an *x* vector.

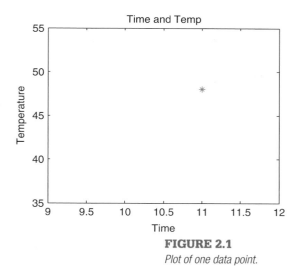

FIGURE 2.1
Plot of one data point.

🌐 **PRACTICE 2.4**

Modify the script **plotonepoint** to prompt the user for the time and temperature, and set the axes based on these values.

FIGURE 2.2
Plot of data points from vectors.

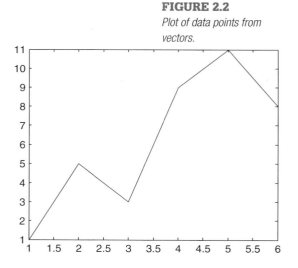

2.5.1.1 *Customizing a Plot: Color, Line Types, Marker Types*

Plots can be done in the Command Window, as shown here, if they are really simple. However, many times it is desirable to customize the plot with labels, titles, and such, so it makes more sense to do this in a script. Using the **help** function for **plot** will show the many options for the line types, colors, and so on. In the script *plotonepoint*, earlier, the string 'r*' specified a red star for the point type. The possible colors are:

```
b        blue
c        cyan
g        green
k        black
m        magenta
r        red
y        yellow
```

The plot symbols, or markers, that can be used are:

```
o        circle
d        diamond
h        hexagram
p        pentagram
+        plus
.        point
s        square
*        star
v        down
         triangle
<        left
         triangle
>        right
         triangle
^        up triangle
x        x-mark
```

Line types can also be specified by the following:

```
--  dashed
-.  dash dot
:  dotted
-  solid
```

If no line type is specified, a solid line is drawn between the points, as seen in the last example.

2.5.2 Simple Related Plot Functions

Other functions that are useful in customizing plots are **clf**, **figure**, **hold**, **legend**, and **grid**. Brief descriptions of these functions are given here; use **help** to find out more about them:

clf clears the Figure Window by removing everything from it.

figure creates a new, empty Figure Window when called without any arguments. Calling it as **figure(n)** where n is an integer is a way of creating and maintaining multiple Figure Windows, and of referring to each individually.

hold is a toggle that freezes the current graph in the Figure Window, so that new plots will be superimposed on the current one. Just **hold** by itself is a *toggle*, so calling this function once turns the hold on, and then the next time turns it off. Alternatively, the commands **hold on** and **hold off** can be used.

legend displays strings passed to it in a legend box in the Figure Window, in order of the plots in the Figure Window.

grid displays grid lines on a graph. Called by itself, it is a toggle that turns the grid lines on and off. Alternatively, the commands **grid on** and **grid off** can be used.

Also, there are many plot types. We will see more in Chapter 10, but another simple plot type is a **bar** chart.

For example, the following script creates two separate Figure Windows. First, it clears the Figure Window. Then, it creates an *x* vector and two different *y* vectors (*y1* and *y2*). In the first Figure Window, it plots the *y1* values using a bar chart. In the second Figure Window, it plots the *y1* values as black lines, puts hold on so that the next graph will be superimposed, and plots the *y2* values as black o's. It also puts a legend on this graph and uses a grid. Labels and titles are omitted in this case since it is generic data.

plot2figs.m

```
% This creates 2 different plots, in 2 different
% Figure Windows, to demonstrate some plot features
clf
x = 1:5; % Not necessary
y1 = [2 11 6 9 3];
y2 = [4 5 8 6 2];
% Put a bar chart in Figure 1
```

(Continued)

```
figure(1)
bar(x,y1)
% Put plots using different y values on one plot
% with a legend
figure(2)
plot(x,y1,'k')
hold on
plot(x,y2,'ko')
grid on
legend('y1','y2')
```

Running this script will produce two separate Figure Windows. If there aren't any other active Figure Windows, the first, which is the bar chart, will be in the one titled in MATLAB Figure 1. The second will be in Figure 2. See Figure 2.3 for both plots.

 PRACTICE 2.5

Modify the script using the **axis** function so that all points are easily seen.

Notice that the first and last points are on the axes, which makes them difficult to see. That is why the **axis** function is frequently used—to create space around the points so that they are all visible.

The ability to pass a vector to a function and have the function evaluate every element of the vector can

A

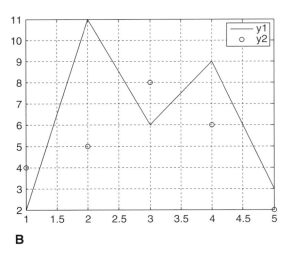

B

FIGURE 2.3

(A) Bar chart produced by script. (B) Plot produced by script, with a grid and legend.

be very useful in creating plots. For example, the following script graphically displays the difference between the **sin** and **cos** functions:

sinncos.m

```
% This script plots sin(x) and cos(x) in the same Figure
% Window for values of x ranging from 0 to 2*pi
clf
x = 0: 2*pi/40: 2*pi;
y = sin(x);
plot(x,y,'ro')
hold on
y = cos(x);
plot(x,y,'b+')
legend('sin', 'cos')
title('sin and cos on one graph')
```

The script creates an *x* vector; iterating through all the values from 0 to 2*π in steps of 2*π/40 gives enough points to get a good graph. It then finds the sine of each x value, and plots these points using red o's. The command **hold on** freezes this in the Figure Window so the next plot will be superimposed. Next, it finds the cosine of each x value and plots these points using blue +'s. The **legend** function creates a legend; the first string is paired with the first plot, and the second string with the second plot. Running this script produces the plot seen in Figure 2.4.

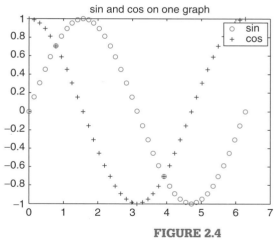

FIGURE 2.4

*Plot of **sin** and **cos** in one Figure Window with a legend.*

2.6 INTRODUCTION TO FILE INPUT/OUTPUT (LOAD **AND** SAVE)

In many cases, input to a script will come from a data file that has been created by another source. Also, it is useful to be able to store output in an external file that can be manipulated and/or printed later. In this section, we will demonstrate how to read from an external data file, and also how to write to an external data file.

 PRACTICE 2.6

Write a script that plots **exp(x)** and **log(x)** for values of x ranging from 0.5 to 2.5.

There are basically three different operations, or *modes*, on files. Files can be:

- Read from
- Written to
- *Appended* to

Writing to a file means writing to a file, from the beginning. Appending to a file is also writing, but starting at the end of the file rather than the beginning. In other words, appending to a file means adding to what was already there.

There are many different file types, which use different filename extensions. For now, we will keep it simple and just work with .dat or .txt files when working with data or text files. There are several methods for reading from files and writing to files; for now we will use the **load** function to read and the **save** function to write to files. More file types and functions for manipulating them will be discussed in Chapter 8.

2.6.1 Writing Data to a File

The **save** function can be used to write data from a matrix to a data file, or to append to a data file. The format is:

```
save filename matrixvariablename -ascii
```

The -ascii qualifier is used when creating a text or data file. The following creates a matrix and then saves the values of the matrix variable to a data file called testfile.dat:

```
>> mymat = rand(2,3)
mymat =
    0.4565  0.8214  0.6154
    0.0185  0.4447  0.7919
>> save testfile.dat mymat -ascii
```

This creates a file called testfile.dat that stores the numbers

```
0.4565  0.8214  0.6154
0.0185  0.4447  0.7919
```

The **type** command can be used to display the contents of the file; notice that scientific notation is used:

```
>> type testfile.dat
4.5646767e-001  8.2140716e-001  6.1543235e-001
1.8503643e-002  4.4470336e-001  7.9193704e-001
```

Note: If the file already exists, the **save** function will overwrite it; **save** always begins writing from the beginning of a file.

2.6.2 Appending Data to a Data File

Once a text file exists, data can be appended to it. The format is the same as previously, with the addition of the qualifier -append. For example, the following creates a new random matrix and appends it to the file just created:

```
>> mymat = rand(3,3)
mymat =

        0.9218          0.4057          0.4103
        0.7382          0.9355          0.8936
        0.1763          0.9169          0.0579
>> save testfile.dat mymat -ascii -append
```

This results in the file testfile.dat containing

```
        0.4565          0.8214          0.6154
        0.0185          0.4447          0.7919
        0.9218          0.4057          0.4103
        0.7382          0.9355          0.8936
        0.1763          0.9169          0.0579
```

Note: Although technically any size matrix could be appended to this data file, in order to be able to read it back into a matrix later there would have to be the same number of values on every row.

2.6.3 Reading from a File

Once a file has been created (as previously), it can be read into a matrix variable. If the file is a data file, the **load** function will read from the file filename.ext (e.g., the extension might be .dat) and create a matrix with the same name as the file. For example, if the data file testfile.dat had been created as shown in the previous section, this would read from it:

 PRACTICE 2.7

Prompt the user for the number of rows and columns of a matrix, create a matrix with that many rows and columns of random numbers, and write it to a file.

```
>> clear
>> load testfile.dat
>> who
Your variables are:
testfile
>> testfile
testfile =

        0.4565          0.8214          0.6154
        0.0185          0.4447          0.7919
        0.9218          0.4057          0.4103
        0.7382          0.9355          0.8936
        0.1763          0.9169          0.0579
```

FIGURE 2.5

Plot of temperature data from a file.

Note: The **load** command works only if there are the same number of values in each line, so that the data can be stored in a matrix, and the **save** command only writes from a matrix to a file. If this is not the case, lower-level file I/O functions must be used; these will be discussed in Chapter 8.

2.6.3.1 *Example: Load from a File and Plot the Data*

As an example, a file called timetemp.dat stores two lines of data. The first line is the times of day, and the second line is the recorded temperature at each of those times. The first value of 0 for the time represents midnight. For example, the contents of the file might be:

0	3	6	9	12	15	18	21
55.5	52.4	52.6	55.7	75.6	77.7	70.3	66.6

The following script loads the data from the file into a matrix called *timetemp*. It then separates the matrix into vectors for the time and temperature, and then plots the data using black + symbols.

timetempprob.m

```
% This reads time and temperature data for an afternoon
% from a file and plots the data
load timetemp.dat

% The times are in the first row, temps in the second row
time = timetemp(1,:);
temp = timetemp(2,:);
% Plot the data and label the plot
plot(time,temp,'k+')
xlabel('Time')
ylabel('Temperature')
title('Temperatures one afternoon')
```

Running the script produces the plot seen in Figure 2.5.

Notice that it is difficult to see the point at time 0 since it falls on the y-axis. The **axis** function could be used to change the axes from the defaults shown here.

PRACTICE 2.8

The sales (in billions) for two separate divisions of the XYZ Corporation for each of the four quarters of 2007 are stored in a file called salesfigs.dat:

```
1.2 1.4 1.8 1.3
2.2 2.5 1.7 2.9
```

1. Create this file (just type the numbers in the editor, and Save As salesfigs.dat).
2. Load the data from the file into a matrix.
3. Write a script that will
   - separate this matrix into two vectors
   - create the plot seen in Figure 2.6 (which uses o's and +'s as the plot symbols):

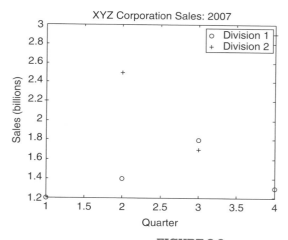

FIGURE 2.6
Plot of sales data from file.

QUICK QUESTION!

Sometimes files are not in the format that is desired. For example, a file expresults.dat has been created that has some experimental results, but the order of the values is reversed in the file:

7	55.2
6	51.9
5	49.5
4	53.4
3	44.3
2	50.0
1	55.5

How could we create a new file that reverses the order?
Answer:
We can **load** from this file into a matrix, use the **flipud** function to "flip" the matrix up-to-down, and then **save** this matrix to a new file:

```
>> load expresults.dat
>> expresults
expresults =
```

7.0000	55.2000
6.0000	51.9000
5.0000	49.5000
4.0000	53.4000
3.0000	44.3000
2.0000	50.0000
1.0000	55.5000

```
>> correctorder = flipud(expresults)
correctorder =
```

1.0000	55.5000
2.0000	50.0000
3.0000	44.3000
4.0000	53.4000
5.0000	49.5000
6.0000	51.9000
7.0000	55.2000

```
>> save neworder.dat correctorder - ascii
```

2.7 USER-DEFINED FUNCTIONS THAT RETURN A SINGLE VALUE

We have already seen the use of many functions in MATLAB. We have used many built-in functions such as **sin**, **fix**, **abs**, **double**, and so on. In this section, *user-defined functions* will be introduced. These are functions that the programmer defines, and then uses, in either the Command Window or in a script.

Functions can return different types of results. For now, we will concentrate on the kind of function that calculates and returns a single result, much like built-in functions such as **sin** and **abs**. Other types of functions will be introduced in Chapter 5.

First, let us review some of what we already know about functions, including the use of built-in functions. Although the use of these functions is straightforward by now, explanations will be given in some detail here in order to compare and contrast the use of user-defined functions.

The **length** function is an example of a built-in function that calculates a single value; it returns the length of a vector. As an example,

```
length(vec)
```

is an expression; it represents the number of elements in the vector *vec*. This expression could be used in the Command Window or in a script. Typically, the value returned from this expression might be assigned to a variable:

```
>> vec = 1:3:10;
>> lv = length(vec)
lv =
    4
```

Alternatively, the length of the vector could be printed

```
>> fprintf('The length of the vector is %d\n', length(vec))
The length of the vector is 4
```

The *call* to the **length** function consists of the name of the function, followed by the *argument* in parentheses. This function takes the argument, and returns a result. What happens when the call to the function is encountered is that *control* is passed to the function itself (in other words, the function begins executing). The argument(s) are also passed to the function. The function executes its statements and does whatever it does (the actual contents of the built-in functions are not generally known or seen by the programmer) in order to determine the number of elements in the vector. Since the function is calculating a single value, this result is then *returned*, and it becomes the value of the expression. Control is also passed back to the expression that called it in the first place, which then continues (e.g., in the first example the value would then be assigned to the variable *lv*, and in the second example the value was printed).

2.7.1 Function Definitions

There are different ways to organize scripts and functions, but for now every function that we write will be stored in a separate M-file, which is why they are commonly called M-file functions.

A function in MATLAB that returns a single result consists of

- The function *header* (the first line); this has

  - the reserved word **function**

  - since the function *returns* a result, the name of the *output argument* followed by the assignment operator =

  - the name of the function (*Important*: This should be the same as the name of the M-file in which this function is stored in order to avoid confusion)

  - the *input arguments* in parentheses; these correspond to the arguments that are passed to the function in the function call

- A comment that describes what the function does (this is printed if **help** is used)

- The *body* of the function, which includes all statements and eventually must assign a value to the output argument

The general form of a *function definition* for a function that calculates and returns one value looks like this:

```
functionname.m
```
```
function outputargument = functionname(input arguments)
% Comment describing the function
Statements here; these must include assigning a value to
    the output argument
```

For example, the following is a function called **calcarea**, which calculates and returns the area of a circle; it is stored in a file called calcarea.m.

```
calcarea.m
```
```
function area = calcarea(rad)
% This function calculates the area of a circle
area = pi * rad * rad;
```

A radius of a circle is passed to the function to the input argument *rad*; the function calculates the area of this circle and stores it in the output argument *area*. In the function header, we have the reserved word **function**, then the output argument *area* followed by the assignment operator =, then the name of the function (the same as the name of the M-file), and then the input argument *rad*, which is the radius. Since there is an output argument in the function header, somewhere in the body of the function we must assign a value to this output argument. This is how a value is returned from the function. In this case, the function is simple and all we have to do is assign to the output argument *area* the value of the built-in constant **pi** multiplied by the square of the input argument *rad*.

The function can be displayed in the Command Window using the **type** command.

```
>> type calcarea
   function area = calcarea(rad)
   % This function calculates the area of a circle
   area = pi * rad * rad;
```

2.7.2 Calling a Function

Here is an example of a call to this function in which the value returned is stored in the default variable *ans*:

```
>> calcarea(4)
ans =
   50.2655
```

Technically, calling the function is done with the name of the file in which the function resides. In order to avoid confusion, it is easiest to give the function the same name as the filename, so that is how it will be presented in this book. In this example, the function name is *calcarea* and the name of the file is calcarea.m. The result returned from this function can also be stored in a variable in an assignment statement; the name could be the same as the name of the output argument in the function itself but that is not necessary; for example, either of these assignments would be fine:

```
>> area = calcarea(5)
area =
   78.5398
>> myarea = calcarea(6)
myarea =
   113.0973
```

The value returned from the *calcarea* function could also be printed using either **disp** or **fprintf**:

```
>> disp(calcarea(4))
   50.2655
>> fprintf('The area is %.1f\n', calcarea(4))
The area is 50.3
```

Notice that the printing is not done in the function itself; rather, the function returns the area and then a print statement can print or display it.

Using **help** with the function displays the contiguous block of comments under the function header:

```
>> help calcarea
   This function calculates the area of a circle
```

2.7.3 Calling a User-Defined Function from a Script

Now, we'll modify our script that prompts the user for the radius and calculates the area of a circle, to call our function *calcarea* to calculate the area of the circle rather than doing this in the script.

```
script3.m
```

```
% This script calculates the area of a circle
% It prompts the user for the radius
radius = input('Please enter the radius:');
% It then calls our function to calculate the
% area and then prints the result
area = calcarea(radius);
fprintf('For a circle with a radius of %.2f,',radius)
fprintf('the area is %.2f\n',area)
```

So, the program consists of the script *script3* and the function *calcarea*.

Running this will produce the following:

```
>> script3
Please enter the radius: 5
For a circle with a radius of 5.00, the area is 78.54
```

2.7.4 Passing Multiple Arguments

In many cases it is necessary to pass more than one argument to a function. For example, the volume of a cone is given by

$$V = \frac{1}{3}\pi r^2 h$$

where *r* is the radius of the circular base and *h* is the height of the cone. Therefore, a function that calculates the volume of a cone needs both the radius and the height:

conevol.m

```
function outarg = conevol(radius, height)
% Calculates the volume of a cone
outarg = (pi/3) * radius * radius * height;
```

Since the function has two input arguments in the function header, two values must be passed to the function when it is called. The order makes a difference. The first value that is passed to the function is stored in the first input argument (in this case, *radius*), and the second argument in the function call is passed to the second input argument in the function header.

 PRACTICE 2.9

Write a script that will prompt the user for the radius and height, call the function *conevol* to calculate the cone volume, and print the result in a nice sentence format. So, the program will consist of a script and the *conevol* function that it calls.

 PRACTICE 2.10

For a project, we need some material to form a rectangle. Write a function *calcrectarea* that will receive the length and width of a rectangle as input arguments, and will return the area of the rectangle. Call the function as shown and store the result in a variable *ra*, and then use the **fprintf** function to print the amount of material required as an integer.

```
>> ra = calcrectarea(3.1, 4.4)
ra =
    13.6400
>> fprintf('need %d sq in\n',...
ceil(ra))
need 14 sq in
```

This is very important: The arguments in the function call must correspond one-to-one with the input arguments in the function header.

Here is an example of calling this function. The result returned from the function is simply stored in the default variable *ans*.

```
>> conevol (4,6.1)
ans =
    102.2065
```

In the next example, the result is printed instead with a format of two decimal places.

```
>> fprintf('The cone volume is %.2f\n',...
conevol(3, 5.5))
The cone volume is 51.84
```

2.7.5 Functions with Local Variables

The functions we've seen so far have been very simple. However, in many cases the calculations in a function are more complicated, and may require the use of extra variables within the function; these are called *local variables*.

For example, a closed cylinder is being constructed of a material that costs a dollar amount per square foot. We will write a function that will calculate and return the cost of the material, rounded up to the nearest square foot, for a cylinder with a given radius and a given height. The total surface area for the closed cylinder is

$$SA = 2\pi rh + 2\pi r^2$$

For example, for a cylinder with a radius of 32 inches and a height of 73 inches, and a cost per square foot of the material of $4.50, the calculation would be given by the following algorithm:

- Calculate the surface area SA = 2 * π * 32 * 73 + 2 * π * 32 * 32 inches squared
- Calculate the SA in square feet = SA/144
- Calculate the total cost = SA in square feet * cost per square foot

The function includes local variables to store the intermediate results.

cylcost.m

```
function outcost = cylcost(radius, height, cost)
% Calculates the cost of constructing a closed
% cylinder

% The radius and height are in inches
% The cost is per square foot
% Calculate surface area in square inches
surf_area = 2 * pi * radius * height + 2 * pi * radius ^ 2;
% Convert surface area in square feet and round up
surf_areasf = ceil(surf_area/144);
% Calculate cost
outcost = surf_areasf * cost;
```

Here are examples of calling the function:

```
>> cylcost(32,73,4.50)
ans =
   661.5000

>> fprintf('The cost would be $%.2f\n', cylcost(32,73,4.50))
The cost would be $661.50
```

SUMMARY

Common Pitfalls

- Putting a space in a variable name
- Spelling a variable name different ways in different places in a script or function

- Forgetting to add the second 's' argument to the **input** function when character input is desired.

- Not using the correct conversion character when printing

- Confusing **fprintf** and **disp**—remember that only **fprintf** can format

Programming Style Guidelines

- Especially for longer scripts and functions, start by writing an algorithm.

- Use comments to document scripts and functions:
  - a block of contiguous comments at the top to describe a script
  - a block of contiguous comments under the function header for functions
  - comments throughout any M-file (script or function) to describe each section

- Make sure that the H1 comment line has useful information.

- Use mnemonic identifier names (names that make sense, e.g., *radius* instead of *xyz*) for variable names and for filenames.

- Make all output easy to read and informative.

- Put a newline character at the end of every string printed by **fprintf** so that the prompt appears on the line below.

- Put informative labels on the x- and y-axes and a title on all plots.

- Keep functions short; normally not longer than one page in length.

- Suppress the output from all assignment statements in a function.

- Functions that return a value do not normally print the value; it should simply be returned by the function.

MATLAB Functions and Commands				
type	plot	axis	legend	load
input	xlabel	clf	grid	save
disp	ylabel	figure	bar	
fprintf	title	hold		

MATLAB Operators
comment %

Exercises

1. Write a simple script that will calculate the volume of a hollow sphere that is ,
   $\frac{4\pi}{3}(r_o^3 - r_i^3)$ where r_i is the inner radius and r_o is the outer radius. Assign a value to a variable for the inner radius, and also assign a value to another variable for the outer radius. Then, using these variables, assign the volume to a third variable. Include comments in the script.

2. The atomic weight is the weight of an atom of a chemical element. For example, the atomic weight of oxygen is 15.9994 and the atomic weight of hydrogen is 1.0079. Write a script that will calculate the molecular weight of hydrogen peroxide, which consists of two atoms of hydrogen and two atoms of oxygen. Include comments in the script. Use **help** to view the comment in your script.

3. Write an **input** statement that will prompt the user for a string. Then, find the length of the string.

4. Write an **input** statement that will prompt the user for a real number, and store it in a variable. Then, use the **fprintf** function to print the value of this variable using two decimal places.

5. The **input** function can be used to enter a vector; for example,

```
>> vec = input('Enter a vector: ')
Enter a vector: 4:7
vec =
     4   5   6   7
```

Experiment with this, and find out how the user can enter a matrix.

6. Experiment, in the Command Window, with using the **fprintf** function for real numbers. Make a note of what happens for each. Use **fprintf** to print the real number 12345.6789

   - without specifying any field width
   - in a field width of 10 with four decimal places
   - in a field width of 10 with two decimal places
   - in a field width of 6 with four decimal places
   - in a field width of 2 with four decimal places

7. Experiment, in the Command Window, with using the **fprintf** function for integers. Make a note of what happens for each. Use fprintf to print the integer 12345

   - without specifying any field width
   - in a field width of 5
   - in a field width of 8
   - in a field width of 3

8. Create the following variables:

```
x = 12.34;
y = 4.56;
```

Then, fill in the **fprintf** statements using these variables that will accomplish the following:

```
>> fprintf(
x is    12.340
>> fprintf(
x is 12
>> fprintf(
y is 4.6
>> fprintf(
y is 4.6      !
```

9. Write a script to prompt the user for the length and width of a rectangle, and print its area with two decimal places. Put comments in the script.

10. Write a script called *echoname* that will prompt the user for his or her name, and then echo print the name in a sentence in the following format (use %s to print it):

```
>> echoname
What is your name? Susan
Wow, your name is Susan!
```

11. Write a script called *echostring* that will prompt the user for a string, and will echo print the string in quotes:

```
>> echostring
Enter your string: hi there
Your string was: 'hi there'
```

12. Write a script that will prompt the user for an angle in degrees. It will then calculate the angle in radians, and then print the result. Note: π radians = 180°.

13. On average, people in a region spend 8 to 10% of their income on food. Write a script that will prompt the user for an annual income. It will then print the range that would typically be spent on food annually. Also, print a monthly range.

14. Wing loading, which is the airplane weight divided by the wing area, is an important design factor in aeronautical engineering. Write a script that will prompt the user for the weight of the aircraft in kilograms, and the wing area in meters squared, and will calculate and print the wing loading of the aircraft in kg/m².

15. Write a script that assigns values for the x-coordinate and then y-coordinate of a point, and then plots this using a green +.

16. Plot **exp(x)** for values of x ranging from –2 to 2 in steps of 0.1. Put an appropriate title on the plot, and label the axes.

17. Create a vector *x* with values ranging from 1 to 100 in steps of 5. Create a vector *y* that is the square root of each value in x. Plot these points. Now, use the **bar** function instead of **plot** to get a bar chart instead.

18. Create a *y* vector that stores random integers in the range from 1 to 100. Create an *x* vector that iterates from 1 to the length of the *y* vector. Experiment with the **plot** function using different colors, line types, and plot symbols.

19. Plot **sin (x)** for x values ranging from 0 to π (in separate Figure Windows):

   - using 10 points in this range
   - using 100 points in this range

20. Atmospheric properties such as temperature, air density, and air pressure are important in aviation. Create a file that stores temperatures in degrees Kelvin at various altitudes. The altitudes are in the first column and the temperatures in the second. For example, it may look like this:

```
1000          288
2000          281
3000          269
5000          256
10000         223
```

   Write a script that will load this data into a matrix, separate it into vectors, and then plot the data with appropriate axis labels and a title.

21. Create a 3 × 6 matrix of random integers, each in the range from 50 to 100. Write this to a file called randfile.dat. Then, create a new matrix of random integers, but this time make it a 2 × 6 matrix of random integers, each in the range from 50 to 100. Append this matrix to the original file. Then, read the file in (which will be to a variable called randfile) just to make sure that it worked!

22. Create a file called testtan.dat that has two lines, with three real numbers on each line (some negative, some positive, in the range from –1 to 3). The file can be created from the editor, or saved from a matrix. Then, **load** the file into a matrix and calculate the tangent of every element in the resulting matrix.

23. A file called hightemp.dat was created some time ago that stores, on every line, a year followed by the high temperature at a specific site for each month of that year. For example, the file might look like this:

```
89    42    49    55    72    63    68    77    82    76    67
90    45    50    56    59    62    68    75    77    75    66
91    44    43    60    60    60    65    69    74    70    70
      etc.
```

   As can be seen, only two digits were used for the year (which was common in the last century). Write a script that will read this file into a matrix, create a new matrix which stores the years correctly as 19xx, and then write this to a new file called y2ktemp.dat. (Hint: Add 1900 to the entire first column of the matrix.) This file, for example, would look like this:

```
1989    42    49    55    72    63    68    77    82    76    67
1990    45    50    56    59    62    68    75    77    75    66
1991    44    43    60    60    60    65    69    74    70    70
        etc.
```

24. Write a function *calcrectarea* that will calculate and return the area of a rectangle. Pass the length and width to the function as input arguments.

25. Write a function *degf_to_K* that will convert a temperature from degrees Fahrenheit (F) to Kelvin (K). It will receive one input argument that is degrees Fahrenheit, and it will return the temperature in K. Here are conversions from F to Celsius and from C to K:

    ```
    C = (F - 32) * 5/9
    K = C + 273.15
    ```

    Here are examples of calling this function:

    ```
    >> ktemp = degf_to_k(33.3)
    ktemp =
       273.8722
    >> fprintf('%.2f degrees F is %.2f degrees K\n', -15.3, ...
    degf_to_k(-15.3))
    -15.30 degrees F is 246.87 degrees K
    ```

26. The hyperbolic sine for an argument x is defined as:

    ```
    hyperbolicsine(x) = (eˣ - e⁻ˣ)/ 2
    ```

    Write a function *hypsin* to implement this. The function should receive one input argument x and return the value of the hyperbolic sine of x. Here are some examples of using the function:

    ```
    >> hypsin(2.1)
    ans =
    4.0219
    >> help hypsin
       Calculates the hyperbolic sine of x
    >> fprintf('The hyperbolic sine of %.1f is %.1f\n',...
    1.9,hypsin(1.9))
    The hyperbolic sine of 1.9 is 3.3
    ```

27. The velocity of an aircraft typically is given in either miles/hour or meters/second. Write a function that will receive one input argument that is the velocity of an airplane in miles per hour and will return the velocity in meters per second. The relevant conversion factors are: one hour = 3600 seconds, one mile = 5280 feet, and one foot = .3048 meters.

28. If a certain amount of money (called the principal P) is invested in a bank account, earning an interest rate i compounded annually, the total amount of money T_n that will be in the account after n years is given by:

    $$T_n = P(1+i)n$$

    Write a function that will receive input arguments for P, i, and n, and will return the total amount of money T_n. Also, give an example of calling the function.

29. The Body Mass Index, or BMI, for a person is defined, using US units, as

$$BMI = 703 * \frac{weight}{height^2}$$

where the weight is in pounds and the height is in total inches. Write a function called *findbmi* that would receive the weight and height as input arguments, and would return the BMI. For example, to find the BMI for a person who weighs 170 pounds and is 5′ 11″ tall (so 71 inches tall), here are two examples of function calls:

```
>> findbmi(170,71)
ans =
    23.7076
>> bmi = findbmi(170,71)
bmi =
    23.7076
```

30. The velocity of a moving fluid can be found from the difference between the total and static pressures P_t and P_s. For water, this is given by $V = 1.016 \sqrt{P_t - P_s}$. Write a function that will receive as input arguments the total and static pressure and will return the velocity of the water.

31. For a project, some biomedical engineering students are designing a device that will monitor a person's heart rate while on a treadmill. The device will let the subject know when the target heart rate has been reached. A simple calculation of the target heart rate (THR) for a moderately active person is

$$THR = (220 - A) * .6$$

where A is the person's age. Write a function that will calculate and return the THR.

32. An approximation for a factorial can be found using Stirling's formula:

$$n! \approx \sqrt{2\pi n} \left(\frac{n}{e} \right)^n$$

Write a function to implement this, passing the value of *n* as an argument.

33. The conversion depends on the temperature and other factors, but an approximation is that 1 inch of rain is equivalent to 6.5 inches of snow. Write a script that prompts the user for the number of inches of rain, calls a function to return the equivalent amount of snow, and prints this result. Write the function also!

34. Write a function called *pickone* that will receive one input argument x, which is a vector, and will return one random element from the vector. For example,

```
>> pickone(4:7)
ans =
    5
>> disp(pickone(-2:0))
```

```
  -1
>> help pickone
pickone(x) returns a random element from vector x
```

35. A function can return a vector as a result. Write a function *vecout* that will receive one integer argument and will return a vector that increments from the value of the input argument to its value plus 5, using the colon operator. For example,

```
>> vecout(4)
ans =
     4   5   6   7   8   9
```

36. If the lengths of two sides of a triangle and the angle between them are known, the length of the third side can be calculated. Given the lengths of two sides (b and c) of a triangle, and the angle between them α in degrees, the third side a is calculated as:

$$a^2 = b^2 + c^2 - 2b\,c\,\cos(\alpha)$$

Write a script *thirdside* that will prompt the user and read in values for b, c, and α (in degrees), and then calculate and print the value of a with three decimal places. (**Note:** To convert an angle from degrees to radians, multiply the angle by $\pi/180$.) The format of the output from the script should look exactly like this:

```
>> thirdside
Enter the first side: 2.2
Enter the second side: 4.4
Enter the angle between them: 50
The third side is 3.429
```

FIGURE 2.7

Plot of Cost and Sales data.

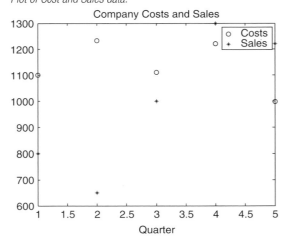

For more practice, write a function to calculate the third side, so the script will call this function.

37. A part is being turned on a lathe. The diameter of the part is supposed to be 20000 mm. The diameter is measured every 10 minutes and the results are stored in a file called partdiam.dat. Create a data file to simulate this. The file will store the time in minutes and the diameter at each time. Plot the data.

38. A file called costssales.dat stores for a company some cost and sales figures for the last n quarters (n is not defined ahead of time). The costs are in the first column, and the sales are in the second column. For example, if five quarters were represented, there would be five lines in the file and it might look like this:

```
1100      800
1233      650
1111     1001
1222     1300
 999     1221
```

Write a script called *salescosts* that will read the data from this file into a matrix. When the script is executed, it will do three things: (1) It will print how many quarters were represented in the file, for example,

```
>> salescosts
There were 5 quarters in the file
```

(2) It will plot the costs using black o's and sales using black *'s in a Figure Window with a legend (using default axes) as seen in Figure 2.7.

(3) It will write the data to a new file called newfile.dat in a different order. The sales will be the first row, and the costs will be the second row. For example, if the file is as shown, the resulting file will store:

```
 800        650        1001       1300       1221
1100       1233        1111       1222        999
```

It should not be assumed that the number of lines in the file is known.

Selection Statements

KEY WORDS

selection statement	action	truth table
branching statement	temporary variable	nesting if-else statement
condition	logical true	cascading if-else
Boolean expression	logical false	statement
relational expression	relational operator	**is** function
logical expression	logical operator	

So far, in the scripts and functions we've seen, every statement was executed in sequence. That is not always desired, and in this section we'll see how to make choices as to whether statements are executed or not, and how to choose between or among them. The statements that accomplish this are called *selection* or *branching* statements.

The MATLAB® software has two basic statements that allow choices: the **if** statement and the **switch** statement. The **if** statement has optional **else** and **elseif** clauses for branching. The **if** statement uses expressions that are logically true or false. These expressions use relational and logical operators. MATLAB also has **is** functions that ask whether something is true or false; these functions will be covered at the end of this chapter.

3.1 RELATIONAL EXPRESSIONS

Conditions in **if** statements use expressions that are conceptually, or logically, either true or false. These expressions are called *relational expressions*, or sometimes *Boolean* or *logical* expressions. These expressions can use both *relational operators*, which relate two expressions of compatible types, and *logical operators*, which operate on logical operands.

The relational operators in MATLAB are:

Operator	Meaning
>	greater than
<	less than
>=	greater than or equals
<=	less than or equals
==	equality
~=	inequality

All concepts should be familiar, although the operators used may be different from those used in other programming languages, or in mathematics classes. In particular, it is important to note that the operator for equality is two consecutive equal signs, not a single equal sign (recall that the single equal sign is the assignment operator).

For numerical operands, the use of these operators is straightforward. For example, 3 < 5 means "3 less than 5," which is conceptually a true expression. However, in MATLAB, as in many programming languages, *logical true* is represented by the integer 1, and *logical false* is represented by the integer 0. So, the expression 3 < 5 actually has the value 1 in MATLAB. Displaying the result of expressions like this in the Command Window demonstrates the values of the expressions.

```
>> 3 < 5
ans =
     1
>> 9 < 2
ans =
     0
```

However, in the Workspace Window, the value shown for the result of these expressions would be true or false. The type of the result is **logical**.

Mathematical operations could be performed on the resulting 1 or 0.

```
>> 5 < 7
ans =
     1
```

```
>> ans + 3
ans =
    4
```

Comparing characters, for example 'a' < 'c', is also possible. Characters are compared using their ASCII equivalent values. So, 'a' < 'c' is conceptually a true expression, because the character 'a' comes before the character 'c'.

```
>> 'a' < 'c'
ans =
    1
```

The logical operators are:

Operator	Meaning
||	or for scalars
&&	and for scalars
~	not

All logical operators operate on logical or Boolean operands. The **not** operator is a unary operator; the others are binary. The **not** operator will take a Boolean expression, which is conceptually true or false, and give the opposite value. For example, ~(3 < 5) is conceptually false since (3 < 5) is true. The **or** operator has two Boolean expressions as operands. The result is true if either or both of the operands are true, and false only if both operands are false. The **and** operator also operates on two Boolean operands. The result of an **and** expression is true only if both operands are true; it is false if either or both are false.

In addition to these logical operators, MATLAB also has a function **xor**, which is the exclusive or function. It returns logical true if one (and only one) of the arguments is true. For example, in the following only the first argument is true, so the result is true:

```
>> xor(3 < 5, 'a' > 'c')
ans =
    1
```

In this example, both arguments are true so the result is false:

```
>> xor(3 < 5, 'a' < 'c')
ans =
    0
```

Given the logical values of true and false in variables x and y, the ***truth table*** (see Table 3.1) shows how the logical operators work for all combinations. Note that the logical operators are commutative (e.g., x || y is the same as y || x).

Table 3.1 Truth Table for Logical Operators

x	y	~x	x ‖ y	x && y	xor(x,y)
true	true	false	true	true	false
true	false	false	true	false	true
false	false	true	false	false	false

As with the numerical operators, it is important to know the operator precedence rules. Table 3.2 shows the rules for the operators that have been covered so far, in the order of precedence.

Table 3.2 Operator Precedence Rules

Operators	Precedence
parentheses ( )	highest
transpose and power', ^	
unary negation (–), not (~)	
multiplication, division *,/,\	
addition, subtraction +, –	
colon operator:	
relational <, <=, >, >=, ==, ~=	
and &&	
or ‖	
assignment =	lowest

QUICK QUESTION!

Assume that there is a variable x that has been initialized. What would the value of the expression

   $3 < x < 5$

be if the value of x is 4? What if the value of x is 7?

Answer: The value of this expression will always be logical true, or 1, regardless of the value of the variable x. Although this is a valid expression, it is probably not what was logically intended. Expressions are evaluated from left to right. So, first the expression $3 < x$ will be evaluated. There are only two possibilities: either this will be conceptually true or false—which means that either the expression will have the value 1 or 0. Then, the rest of the expression will be evaluated, which will be either $1 < 5$ or $0 < 5$. Both of these expressions are conceptually true. So, the value of x does not matter: the expression $3 < x < 5$ would be true regardless of the value of the variable x. This is a logical error; it would not enforce the desired range. If we wanted an expression that was logically true only if x was in the range from 3 to 5, we could write $3 < x$ && $x < 5$.

3.2 THE IF STATEMENT

The **if** statement chooses whether or not another statement, or group of statements, is executed. The general form of the **if** statement is:

```
if condition
    action
end
```

A *condition* is a relational expression that is conceptually, or logically, either true or false. The *action* is a statement, or a group of statements, that will be executed if the condition is true. When the **if** statement is executed, first the condition is evaluated. If the value of the condition is conceptually true, the action will be executed, and if not, the action will not be executed. The action can be any number of statements until the reserved word **end**; the action is naturally bracketed by the reserved words **if** and **end**. (Note: This is different from the **end** that is used as an index into a vector or matrix.)

 PRACTICE 3.1

Think about what would be produced by the following expressions, and then type them to verify your answers.

```
4 > 3 + 1
'e' == 'd' + 1
3 < 9 - 2
(3 < 9) - 2
4 == 3 + 1 && 'd' > 'c'
3 >= 2 || 'x' == 'y'
xor(3 >= 2,'x' == 'y')
xor(3 >= 2,'x' ~= 'y')
```

For example, the following **if** statement checks to see whether the value of a variable is negative. If it is, the value is changed to a positive number by using the absolute value function; otherwise nothing is changed.

```
if num < 0
   num = abs(num)
end
```

If statements can be entered in the Command Window, although they generally make more sense in scripts or functions. In the Command Window, the **if** line would be entered, then the Enter key, then the action, the Enter key, and finally **end** and Enter; the results will immediately follow. For example, the previous **if** statement is shown twice here. Notice that the output from the assignment is not suppressed, so the result of the action will be shown if the action is executed. The first time the value of the variable is negative so the action is executed and the variable is modified, but in the second case the variable is positive so the action is skipped.

```
>> num = -4;
>> if num < 0
   num = abs(num)
end
num =
     4
>> num = 5;
>> if num < 0
   num = abs(num)
end
>>
```

This may be used, for example, to make sure that the square root function is not used on a negative number. The following script prompts the user for a number, and prints the square root. If the user enters a negative number, the **if** statement changes it to positive before taking the square root.

sqrtifexamp.m

```
% Prompt the user for a number and print its sqrt
num = input('Please enter a number: ');
% If the user entered a negative number, change it
if num < 0
    num = abs(num);
end
fprintf('The sqrt of %.1f is %.1f\n',num,sqrt(num))
```

Here are two examples of running this script:

```
>> sqrtifexamp
Please enter a number: −4.2
The sqrt of 4.2 is 2.0

>> sqrtifexamp
Please enter a number: 1.44
The sqrt of 1.4 is 1.2
```

In this case, the action of the **if** statement was a single assignment statement. The action can be any number of valid statements. For example, we may wish to print a note to the user to say that the number entered was being changed.

sqrtifexampii.m

```
% Prompt the user for a number and print its sqrt
num = input('Please enter a number: ');
% If the user entered a negative number, tell
% the user and change it
if num < 0
    disp('OK, we''ll use the absolute value')
    num = abs(num);
end
fprintf('The sqrt of %.1f is %.1f\n',num,sqrt(num))
```

```
>> sqrtifexampii
Please enter a number: -25
OK, we'll use the absolute value
The sqrt of 25.0 is 5.0
```

Notice the use of two single quotes in the **disp** statement in order to print one single quote.

PRACTICE 3.2

Write an **if** statement that would print "No overtime for you!" if the value of a variable *hours* is less than 40. Test the **if** statement for values of *hours* both less than and greater than 40.

QUICK QUESTION!

Let's say that we want to create a vector of increasing values from *mymin* to *mymax*. We will write a function *createvec* that will receive two input arguments, *mymin* and *mymax*, and will return a vector with values from *mymin* to *mymax* in steps of 1. First, we would make sure that the value of *mymin* is less than the value of *mymax*. If not, we would need to exchange their values before creating the vector. How would we accomplish this?

Answer: To exchange values, a third, temporary variable is required. For example, let's say that we have two variables *a* and *b*, storing the values:

```
a = 3;
b = 5;
```

To exchange values, we could not just assign the value of *b* to *a*, as follows:

```
a = b;
```

If that was done, then the value of *a* (3), is lost! Instead, we need to assign the value of *a* first to a **temporary variable** so that the value is not lost. The algorithm would be:

- Assign the value of *a* to *temp*
- Assign the value of *b* to *a*
- Assign the value of *temp* to *b*

```
>> temp = a;
>> a = b
a =
     5
>> b = temp
b =
     3
```

Now, for the function: An **if** statement is used to determine whether or not the exchange is necessary.

createvec.m

```
function outvec = createvec(mymin, mymax)
% Create a vector that iterates from mymin to mymax

% If the minimum isn't smaller than the maximum,
% exchange the values using a temporary variable
if mymin > mymax
    temp = mymin;
    mymin = mymax;
    mymax = temp;
end
% Use the colon operator to create the vector
outvec = mymin:mymax;
```

(Continued)

Here are examples of calling the function:

```
>> createvec(4,6)
ans =
     4     5     6
```

```
>> createvec(7,3)
ans =
     3     4     5     6     7
```

3.2.1 Representing Logical True and False

It has been stated that expressions that are conceptually true actually have the integer value of 1, and expressions that are conceptually false actually have the integer value of 0. Representing the concepts of logical true and false in MATLAB is slightly different: the concept of false is represented by the integer value of 0, but the concept of true can be represented by any nonzero value (not just the integer 1). This can lead to some strange Boolean expressions. For example, consider the following **if** statement:

```
>> if 5
    disp('Yes, this is true!')
   end
Yes, this is true!
```

Since 5 is a nonzero value, it is a way of saying true. Therefore, when this Boolean expression is evaluated, it will be true, so the **disp** function will be executed and "Yes, this is true" is displayed. Of course, this is a pretty bizarre **if** statement, one that hopefully would not ever be encountered!

However, a simple mistake in an expression can lead to this kind of result. For example, let's say that the user is prompted for a choice of Y or N for a yes/no question:

```
letter = input('Choice (Y/N): ','s');
```

In a script we might want to execute a particular action if the user responded with 'Y'. Most scripts would allow the user to enter either lower- or uppercase (e.g., either 'y' or 'Y') to indicate yes. The proper expression that would return true if the value of letter was 'y' or 'Y' would be

```
letter == 'y' || letter == 'Y'
```

However, if by mistake this was written as:

```
letter == 'y' || 'Y'
```

this expression would *always* be true, regardless of the value of the variable *letter*. This is because 'Y' is a nonzero value, so it is a true expression. The first part of the expression may be false, but since the second expression is true the entire expression would be true.

3.3 THE IF-ELSE STATEMENT

The **if** statement chooses whether an action is executed or not. Choosing between two actions, or choosing from several actions, is accomplished using **if-else**, nested **if**, and **switch** statements.

The **if-else** statement is used to choose between two statements, or sets of statements. The general form is:

```
if condition
    action1
else
    action2
end
```

First, the condition is evaluated. If it is conceptually true, then the set of statements designated as action1 is executed, and that is it for the **if-else** statement. If instead the condition is conceptually false, the second set of statements designated as action2 is executed, and that's it. The first set of statements is called the action of the **if** clause; it is what will be executed if the expression is true. The second set of statements is called the action of the **else** clause; it is what will be executed if the expression is false. One of these actions, and only one, will be executed—which one depends on the value of the condition.

For example, to determine and print whether or not a random number in the range from 0 to 1 is less than 0.5, an **if-else** statement could be used:

```
if rand < 0.5
    disp('It was less than .5!')
else
    disp('It was not less than .5!')
end
```

One application of an **if-else** statement is to check for errors in the inputs to a script. For example, an earlier script prompted the user for a radius, and then used that to calculate the area of a circle. However, it did not check to make sure that the radius was valid (e.g., a positive number). Here is a modified script that checks the radius:

 PRACTICE 3.3

Write a script *printsindegorrad* that will:

- Prompt the user for an angle.
- Prompt the user for (r)adians or (d)egrees, with radians as the default.
- Use the **sind** function to get the sine of the angle in degrees if the user entered 'd'; otherwise, use the **sin** function. Which sine function to use will be based solely on whether the user entered a 'd' or not: a 'd' means degrees, so sind is used; otherwise, for any other character the default of radians is assumed so **sin** is used.
- Print the result.

Here are examples of running the script:

```
>> printsindegorrad
Enter the angle: 45
(r)adians (the default) or
(d)egrees: d
The sin is 0.71

>> printsindegorrad
Enter the angle: pi
(r)adians (the default) or
(d)egrees: r
The sin is 0.00
```

checkradius.m

```
% This script calculates the area of a circle
% It error-checks the user's radius
radius = input('Please enter the radius: ');
if radius <= 0
    fprintf('Sorry; %.2f is not a valid radius\n',radius)
else
    area = calcarea(radius);
    fprintf('For a circle with a radius of %.2f,',radius)
    fprintf('the area is %.2f\n',area)
end
```

Examples of running this script when the user enters invalid and then valid radii are shown here:

```
>> checkradius
Please enter the radius: -4
Sorry; -4.00 is not a valid radius

>> checkradius
Please enter the radius: 5.5
For a circle with a radius of 5.50, the area is 95.03
```

The **if-else** statement in this example chooses between two actions: printing an error message, or actually using the radius to calculate the area, and then printing out the result. Notice that the action of the **if** clause is a single statement, whereas the action of the **else** clause is a group of three statements.

3.4 NESTED IF-ELSE STATEMENTS

The **if-else** statement is used to choose between two statements. In order to choose from more than two statements, the **if-else** statements can be nested, one inside of another. For example, consider implementing the following continuous mathematical function $y = f(x)$:

```
y = 1 for x < -1
y = x² for -1 ≤ x ≤ 2
y = 4 for x > 2
```

The value of y is based on the value of x, which could be in one of three possible ranges. Choosing which range could be accomplished with three separate **if** statements, as follows:

```
if x < -1
    y = 1;
end
if x > = -1 && x < = 2
    y = x^2;
end
if x > 2
    y = 4;
end
```

Since the three possibilities are mutually exclusive, the value of y can be determined by using three separate **if** statements. However, this is not very efficient code: all three Boolean expressions must be evaluated, regardless of the range in which x falls. For example, if x is less than -1, the first expression is true and 1 would be assigned to y. However, the two expressions in the next two **if** statements are still evaluated. Instead of writing it this way, the expressions can be *nested* so that the statement ends when an expression is found to be true:

```
if x < -1
    y = 1;
else
    % If we are here, x must be > = -1
    % Use an if-else statement to choose
    % between the two remaining ranges
    if x > = -1 && x < = 2
        y = x^2;
    else
        % No need to check
        % If we are here, x must be > 2
        y = 4;
    end
end
```

By using a nested **if-else** to choose from among the three possibilities, not all conditions must be tested as they were in the previous example. In this case, if x is less than -1, the statement to assign 1 to y is executed, and the if-else statement is completed so no other conditions are tested. If, however, x is not less than -1, then the else clause is executed. If the else clause is executed, then we already know that x is greater than or equal to -1 so that part does not need to be tested. Instead, there are only two remaining possibilities: either x is less than or equal to 2, or it is greater than 2. An **if-else** statement is used to choose between those two possibilities. So, the action of the **else** clause was another **if-else** statement. Although it is long, this is one **if-else** statement, a nested **if-else** statement. The actions are indented to show the structure.

Nesting **if-else** statements in this way can be used to choose from among three, four, five, six, or more options—the possibilities are practically endless!

This is actually an example of a particular kind of nested **if-else** called a cascading **if-else** statement. In this type of nested **if-else** statement, the conditions and actions cascade in a stair-like pattern.

The Elseif **Clause**

The Programming Concept

In most programming languages, choosing from multiple options means using nested **if-else** statements. However, MATLAB has another method of accomplishing this, using the **elseif** clause.

The Efficient Method

To choose from among more than two actions, the **elseif** clause is used. For example, if there are n choices (where $n > 3$ in this example), the following general form would be used:

```
if condition1
    action1
elseif condition2
    action2
elseif condition3
    action3
% etc: there can be many of these
else
    actionn % the nth action
end
```

The actions of the **if**, **elseif**, and **else** clauses are naturally bracketed by the reserved words **if**, **elseif**, **else**, and **end**.

For example, the previous example could be written using the **elseif** clause rather than nesting **if-else** statements:

```
if x < -1
    y = 1;
elseif x >= -1 && x <= 2
    y = x^2;
else
    y = 4;
end
```

So, there are three ways of accomplishing this task: using three separate **if** statements, using nested **if-else** statements, and using an **if** statement with **elseif** clauses, which is the simplest.

This could be implemented in a function that receives a value of x and returns the corresponding value of y:

calcy.m

```
function y = calcy(x)
% This function calculates y based on x:
% y = 1 for x < -1
% y = x² for -1 ≤ x ≤ 2
% y = 4 for x > 2
if x < -1
    y = 1;
elseif x >= -1 && x <=2
    y = x^2;
else
    y = 4;
end
```

```
>> x = 1.1;
>> y = calcy(x)
y =
    1.2100
```

QUICK QUESTION!

How could you write a function to determine whether an input argument is a scalar, a vector, or a matrix?

Answer: To do this, the **size** function can be used on the input argument. If both the number of rows and columns is equal to 1, then the input argument is a scalar. If, on the other hand, only one dimension is 1, the input argument is a vector (either a row or column vector). If neither dimension is 1, the input argument is a matrix. These three options can be tested using an **if-else** statement. In this example, the word 'scalar', 'vector', or 'matrix' is returned from the function.

findargtype.m

```
function outtype = findargtype(inputarg)
% This function determines whether the input
% argument is a scalar, vector, or matrix
[r c] = size(inputarg);
if r == 1 && c == 1
```

(Continued)

```
            outtype = 'scalar';
    elseif r == 1 || c == 1
            outtype = 'vector';
    else
            outtype = 'matrix';
    end
```

Notice that there is no need to check for the last case: if the input argument isn't a scalar or a vector, it must be a matrix!

Here are examples of calling this function:

```
>> findargtype(33)
ans =
scalar
```

```
>> findargtype(2:5)
ans =
vector
>> findargtype(randint(2,3))
ans =
matrix
```

 PRACTICE 3.4

Modify the function *findargtype* to return either 'scalar', 'row vector', 'column vector', or 'matrix', depending on the input argument.

Another example demonstrates choosing from more than just a few options. The following function receives an integer quiz grade, which should be in the range from 0 to 10. The program then returns a corresponding letter grade, according to the following scheme: a 9 or 10 is an 'A', an 8 is a 'B', a 7 is a 'C', a 6 is a 'D', and anything below that is an 'F'. Since the possibilities are mutually exclusive, we could implement the grading scheme using separate **if** statements. However, it is more efficient to have one **if-else** statement with multiple **elseif** clauses. Also, the function returns the value 'X' if the quiz grade is not valid. The function does assume that the input is an integer.

letgrade.m

```
function grade = letgrade(quiz)
% This function returns the letter grade corresponding
% to the integer quiz grade argument

% First, error-check
if quiz < 0 || quiz > 10
    grade = 'X';
% If here, it is valid so figure out the
% corresponding letter grade
elseif quiz == 9 || quiz == 10
```

(Continued)

```
      grade = 'A';
  elseif quiz == 8
      grade = 'B';
  elseif quiz == 7
      grade = 'C';
  elseif quiz == 6
      grade = 'D';
  else
      grade = 'F';
  end
```

Here are three examples of calling this function:

```
>> quiz = 8;
>> lettergrade = letgrade(quiz)
lettergrade =
B

>> quiz = 4;
>> letgrade(quiz)
ans =
F

>> quiz = 22;
>> lg = letgrade(quiz)
lg =
X
```

In the part of this **if** statement that chooses the appropriate letter grade to return, all the Boolean expressions are testing the value of the variable *quiz* to see if it is equal to several possible values, in sequence (first 9 or 10, then 8, then 7, etc.). This part can be replaced by a **switch** statement.

3.5 THE SWITCH STATEMENT

A **switch** statement can often be used in place of a nested **if-else** or an **if** statement with many **elseif** clauses. Switch statements are used when an expression is tested to see whether it is *equal to* one of several possible values.

The general form of the **switch** statement is:

```
switch switch_expression
  case caseexp1
    action1
```

```
   case caseexp2
     action2
   case caseexp3
     action3
   % etc: there can be many of these
   otherwise
     actionn
end
```

The **switch** statement starts with the reserved word **switch**, and ends with the reserved word **end**. The switch_expression is compared, in sequence, to the case expressions (caseexp1, caseexp2, etc.). If the value of the switch_ expression matches caseexp1, for example, then action1 is executed and the **switch** statement ends. If the value matches caseexp3, then action3 is executed, and in general if the value matches caseexpi, where i can be any integer from 1 to n, then actioni is executed. If the value of the switch_ expression does not match any of the case expressions, the action after the word **otherwise** is executed.

For the previous example, the **switch** statement can be used as follows:

switchletgrade.m

```
function grade = switchletgrade(quiz)
% This function returns the letter grade corresponding
% to the integer quiz grade argument using switch

% First, error-check
if quiz < 0 || quiz > 10
    grade = 'X';
else
    % If here, it is valid so figure out the
    % corresponding letter grade using a switch
    switch quiz
      case 10
          grade = 'A';
      case 9
          grade = 'A';
      case 8
          grade = 'B';
```

(Continued)

```
        case 7
            grade = 'C';
        case 6
            grade = 'D';
        otherwise
            grade = 'F';
    end
end
```

Here are two examples of calling this function:

```
>> quiz = 22;
>> lg = switchletgrade(quiz)
lg =
X

>> quiz = 9;
>> switchletgrade(quiz)
ans =
A
```

Note that it is assumed that the user will enter an integer value. If the user does not, either an error message will be printed or an incorrect result will be returned. Methods for remedying this will be discussed in Chapter 4.

Since the same action of printing 'A' is desired for more than one case, these can be combined as follows:

```
switch quiz
    case {10,9}
        grade = 'A';
    case 8
        grade = 'B';
        % etc.
```

(The curly braces around the case expressions 10 and 9 are necessary.)

In this example, we error-checked first using an **if-else** statement, and then if the grade was in the valid range, used a **switch** statement to find the corresponding letter grade.

Sometimes the **otherwise** clause is used instead for the error message. For example, if the user is supposed to enter only a 1, 3, or 5, the script might be organized as follows:

```
switcherror.m
```

```
% Example of otherwise for error message
choice = input('Enter a 1, 3, or 5: ');
switch choice
    case 1
        disp('It''s a one!!')
    case 3
        disp('It''s a three!!')
    case 5
        disp('It''s a five!!')
    otherwise
        disp('Follow directions next time!!')
end
```

In this case, actions are taken if the user correctly enters one of the valid options. If the user does not, the **otherwise** clause handles printing an error message. Note the use of two single quotes within the string to print one.

```
>> switcherror
Enter a 1, 3, or 5: 4
Follow directions next time!!
```

3.6 THE MENU FUNCTION

MATLAB also has a built-in function called **menu** that will display a Figure Window with push buttons for the choices. The first string passed to the **menu** function is the heading, and the rest are labels that appear on the push buttons. The function returns the number of the button that is pressed. For example,

```
>> mypick = menu('Pick a pizza','Cheese','Shroom','Sausage');
```

will display the Figure Window seen in Figure 3.1 and store the result of the user's button press in the variable *mypick*.

There are three buttons, whose equivalent values are 1, 2, and 3. For example, if the user presses the Sausage button, mypick would have the value 3:

```
>> mypick
mypick =
    3
```

FIGURE 3.1
Menu Figure Window.

Note that the strings 'Cheese', 'Shroom', and 'Sausage' are just labels on the buttons. The actual value of the button press in this case would be 1, 2, or 3.

A script that uses this **menu** function then would use either an **if-else** statement or a **switch** statement to take an action based on the button pressed. For example, the following script simply prints which pizza to order, using a **switch** statement.

pickpizza.m

```
%This script asks the user for a type of pizza
% and prints which type to order using a switch
mypick = menu('Pick a pizza','Cheese','Shroom','Sausage');
switch mypick
   case 1
      disp('Order a cheese pizza')
   case 2
      disp('Order a mushroom pizza')
   case 3
      disp('Order a sausage pizza')
   otherwise
      disp('No pizza for us today')
end
```

Here is an example of running this script and clicking the Sausage button:

```
>> pickpizza
Order a sausage pizza
```

QUICK QUESTION!

How could the **otherwise** action get executed in this **switch** statement?

Answer: If the user clicks the red X on the top-right of the menu box instead of one of the three buttons, the value returned from the menu function will be 0, which will cause the **otherwise** clause to be executed:

```
>> pickpizza
No pizza for us today
```

Instead of using a **switch** statement in this script, an alternative method would be to use an **if-else** statement with **elseif** clauses.

pickpizzaifelse.m

```
%This script asks the user for a type of pizza
% and prints which type to order using if-else
mypick = menu('Pick a pizza','Cheese','Shroom','Sausage');
if mypick == 1
    disp('Order a cheese pizza')
elseif mypick == 2
    disp('Order a mushroom pizza')
elseif mypick == 3
    disp('Order a sausage pizza')
else
    disp('No pizza for us today')
end
```

 PRACTICE 3.5

Write a function that will receive one number as an input argument. It will use the **menu** function to display 'Choose a function' and will have buttons labeled 'fix', 'floor', and 'abs'. Using a **switch** statement, the function will then calculate and return the requested function (e.g., if 'abs' is chosen, the function will return the absolute value of the input argument). Choose a fourth function to return if the user does not push a button, as described earlier.

3.7 THE IS FUNCTIONS IN MATLAB

There are a lot of functions that are built into MATLAB that test whether or not something is true; these function names begin with the word **is**. Since these functions are frequently used in **if** statements, they are introduced in this chapter. For example, the function called **isletter** returns logical 1 if the argument is a letter of the alphabet, or 0 if it is not:

```
>> isletter('h')
ans =
    1
>> isletter('4')
ans =
    0
```

This result could also be found by comparing a character's position within the character encoding, for example, for lower case letters.

The Programming Concept

```
mychar >= 'a' && mychar <= 'z'
```

The Efficient Method

```
isletter(mychar)
```

The function **isempty** returns logical true if a variable is empty, and logical false if it has a value, or an error message if the variable does not exist. Therefore, it can be used to determine whether a variable has a value yet or not. For example,

```
>> clear
>> isempty(evec)
??? Undefined function or variable 'evec'.

>> evec = [];
>> isempty(evec)
ans =
    1

>> evec = [evec 11];
>> isempty(evec)
ans =
    0
```

The **isempty** function will also determine whether or not a string variable is empty. For example, this function can be used to determine whether the user entered a string in the **input** function:

```
>> istr = input('Please enter a string: ','s');
Please enter a string:
>> isempty(istr)
ans =
    1
```

The function **iskeyword** will determine whether or not a name is a keyword in MATLAB, and therefore something that cannot be used as an identifier name. By itself (with no arguments), it will return the list of all keywords. Notice that the names of functions like **sin** are not keywords, so their values can be overwritten if used as an identifier name.

 PRACTICE 3.6

Prompt the user for a string, and then print either the string that the user entered or an error message if the user did not enter anything.

```
>> iskeyword('sin')
ans =
    0

>> iskeyword('switch')
ans =
    1

>> iskeyword
ans =
    'break'
```

```
'case'
'catch'
% etc.
```

There are many other **is** functions; the list can be found in the Help browser.

SUMMARY

Common Pitfalls

- Using = instead of == for equality

- Not using quotes when comparing a string variable to a string; for example,

```
letter == y
```

instead of

```
letter == 'y'
```

- Confusing && and ||

- Confusing || and **xor**

- Putting a space in two-character operators; for example, typing < = instead of <=

- Not spelling out an entire Boolean expression; for example, typing

```
radius || height <= 0
```

instead of

```
radius <= 0 || height <= 0
```

or typing

```
letter == 'y' || 'Y'
```

instead of

```
letter == 'y' || letter == 'Y'
```

Note that these are logically incorrect, but would not result in error messages. Note also that the expression letter == 'y' || 'Y' will *always* be true, regardless of the value of the variable *letter*, since 'Y' is a nonzero value and therefore a true expression.

- Using an **if** statement instead of an **if-else** statement for error-checking; for example,

```
if error occurs
    print error message
end
continue rest of program
```

instead of

```
if error occurs
    print error message
else
    continue rest of program
end
```

In the first example, the error message would be printed but then the program would continue anyway.

Programming Style Guidelines

- Use indentation to show the structure of a script or function. In particular, the actions in an **if** statement should be indented.

- When the **else** clause isn't needed, use an **if** statement rather than an **if-else**. For example,

```
if unit == 'i'
    len = len * 2.54;
else
    len = len; % this does nothing so skip it!
end
```

Instead, just use:

```
if unit == 'i'
    len = len * 2.54;
end
```

- Do not put unnecessary conditions on **else** or **elseif** clauses. For example, the following prints one thing if the value of a variable *number* is equal to 5, and something else if it is not.

```
if number == 5
    disp('It is a 5')
elseif number ~= 5
    disp ('It is not a 5')
end
```

The second condition, however, is not necessary. Either the value is 5 or not, so just the **else** would handle this:

```
if number == 5
    disp('It is a 5')
else
    disp ('It is not a 5')
end
```

- When using the **menu** function, make sure that the possibility that the user clicks the red X on the menu box rather than pushing one of the buttons is handled.

MATLAB Functions and Commands

xor	isletter	if	switch
sind	isempty	if-else	elseif clause
menu	iskeyword		

MATLAB Operators

less than <	greater than or equals >=	inequality ~= or for scalars ||	and for scalars && not ~
greater than >	equality ==		
less than or equals <=			

Exercises

1. What would be the result of the following expressions?

```
'b' >= 'c' - 1
3 == 2 + 1
(3 == 2) + 1
xor(5 < 6, 8 > 4)
```

2. Write a script that tests whether the user can follow instructions. It prompts the user to enter an x. If the user enters anything other than an x, it prints an error message; otherwise, the script does nothing.

3. Write a function *nexthour* that will receive one integer argument, which is an hour of the day, and will return the next hour. This assumes a 12-hour clock, so for example the next hour after 12 would be 1. Here are two examples of calling this function.

```
>> fprintf('The next hour will be %d.\n',nexthour(3))
The next hour will be 4.
>> fprintf('The next hour will be %d.\n',nexthour(12))
The next hour will be 1.
```

4. Write a script to calculate the volume of a pyramid, which is 1/3 * base * height, where the base is length * width. Prompt the user to enter values for the length, width, and the height and then calculate the volume of the pyramid. When the user

enters each value, he or she will then be prompted also for either i for inches, or c
for centimeters. (Note: 2.54 cm = 1 inch). The script should print the volume in cubic
inches with three decimal places. As an example, the format will be:

```
This program will calculate the volume of a pyramid.
Enter the length of the base: 50
Is that i or c? i
Enter the width of the base: 6
Is that i or c? c
Enter the height: 4
Is that i or c? i
The volume of the pyramid is xxx.xxx cubic inches.
```

5. Write a script that will ask the user "Are you an engineer? (Y/N)" and then print an
   appropriate response based on the user's answer. Decide what is appropriate!

6. Write a script to prompt the user for a character, and then print whether or not it is
   a letter of the alphabet.

7. Write a script that will prompt the user for a numerator and a denominator for a
   fraction. If the denominator is 0, it will print an error message saying that division
   by 0 is not possible. If the denominator is not 0, it will print the result of the fraction.

8. The systolic and diastolic blood pressure readings are found when the heart is
   pumping and the heart is at rest, respectively. A biomedical experiment is being
   conducted on only subjects whose blood pressure is optimal. This is defined as a
   systolic blood pressure less than 120 and a diastolic blood pressure less than 80.
   Write a script that will prompt for the systolic and diastolic blood pressures of a
   person, and will print whether or not that person is a candidate for this experiment.

9. The continuity equation in fluid dynamics for steady fluid flow through a stream
   tube equates the product of the density, velocity, and area at two points that have
   varying cross-sectional areas. For incompressible flow, the densities are constant
   so the equation is $A_1V_1 = A_2V_2$. If the areas and V_1 are known, V_2 can be found as
   $\frac{A_1}{A_2} V_1$. Therefore, whether the velocity at the second point increases or decreases
   depends on the areas at the two points. Write a script that will prompt the user
   for the two areas in feet squared, and will print whether the velocity at the second
   point will increase, decrease, or remain the same as at the first point.

10. Write a function *createvec_m_to_n* that will create and return a vector of integers
    from m to n (where m is the first input argument and n is the second), regardless of
    whether m is less than n or greater than n. If m is equal to n, the vector will just be
    1 × 1 or a scalar. Here are some examples of calling the function:

```
>> createvec_m_to_n(8,5)
ans =
     8  7  6  5
>> createvec_m_to_n(6,6)
ans =
     6
```

```
>> result = createvec_m_to_n(4,5)
result =
       4  5
>> help createvec_m_to_n
Creates a vector of integers from m to n
```

11. In a script, the user is supposed to enter either a 'y' or 'n' in response to a prompt. The user's input is read into a character variable called *letter*. The script will print "OK, continuing" if the user enters either a 'y' or 'Y' or it will print "OK, halting" if the user enters an 'n' or 'N', or it will print "Error" if the user enters anything else. Put this statement in the script first:

```
letter = input('Enter your answer: ', 's');
```

    Write the script using a single nested **if-else** statement (**elseif** clause is permitted).

12. Write the script from the previous exercise using a **switch** statement instead.

13. In aerodynamics, the Mach number is a critical quantity. It is defined as the ratio of the speed of an object (such as an aircraft) to the speed of sound. If the Mach number is less than one, the flow is subsonic. If the Mach number is equal to one, the flow is transonic. If the Mach number is greater than one, the flow is supersonic. Write a script that will prompt the user for the speed of an aircraft and the speed of sound at the aircraft's current altitude and will print whether the condition is subsonic, transonic, or supersonic.

14. Write a script that will prompt the user for a temperature in degrees Celsius, and then an F for Fahrenheit or K for Kelvin. The script will print the corresponding temperature in the scale specified by the user. For example, the output might look like this:

```
Enter the temp in degrees C: 29.3
Do you want F or K? F
The temp in degrees F is 84.7
```

    The format of the output should be exactly as specified here. The conversions are:

$$F = \frac{9}{5}C + 32$$

```
K = C+273.15
```

15. Write a function *isdivby4* that will receive an integer input argument, and will return logical 1 for true if the input argument is divisible by 4, or logical false if it is not.

16. Write a script that will generate one random integer, and will print whether the random integer is an even or an odd number. Hint: An even number is divisible by 2, whereas an odd number is not, so check the remainder after dividing by 2.

17. Write a function *isint* that will receive a number input argument *innum*, and will return 1 for true if this number is an integer, or 0 for false if not. Use the fact that *innum* should be equal to **int32(innum)** if it is an integer. Unfortunately, due to round-off errors, it is possible to get 1 for true if the input argument is close to an integer. For example,

```
>> isint(4)
ans =
     1
>> isint(4.9999)
ans =
     0
>> isint(4.999999999999999999999999999)
ans =
     1
```

18. A Pythagorean triple is a set of positive integers (a,b,c) such that $a^2 + b^2 = c^2$. Write a function *ispythag* that will receive three positive integers (a, b, c in that order) and will return 1 for true if they form a Pythagorean triple, or 0 for false if not.

19. Whether a storm is a tropical depression, tropical storm, or hurricane is determined by the average sustained wind speed. In miles per hour, a storm is a tropical depression if the winds are less than 38 mph. It is a tropical storm if the winds are between 39 and 73 mph, and it is a hurricane if the wind speeds are $>= 74$ mph. Write a script that will prompt the user for the wind speed of the storm, and will print which type of storm it is.

20. Hurricanes are categorized based on wind speeds. The following table shows the Category number for hurricanes with varying wind ranges and what the storm surge is (in feet above normal).

1	74-95	4-5
2	96-110	6-8
3	111-130	9-12
4	131-155	13-18
5	>155	>18

    Write a script that will prompt the user for the wind speed, and will print the hurricane category number and the typical storm surge.

21. Clouds are generally classified as high, middle, or low level. The height of the cloud is the determining factor, but the ranges vary depending on the temperature. For example, in tropical regions the classifications may be based on the following height ranges (given in feet):

low	0-6500
middle	6500-20000
high	> 20000

    Write a script that will prompt the user for the height of the cloud in feet, and print the classification.

22. Rewrite the following **switch** statement as one nested **if-else** statement (**elseif** clauses may be used). Assume that there is a variable *letter* and that it has been initialized.

```
switch letter
        case 'x'
            disp('Hello')
        case {'y', 'Y'}
            disp('Yes')
        case 'Q'
            disp('Quit')
        otherwise
            disp('Error')
end
```

23. Rewrite the following nested **if-else** statement as a **switch** statement that accomplishes exactly the same thing. Assume that num is an integer variable that has been initialized, and that there are functions *f1*, *f2*, *f3*, and *f4*. Do not use any **if** or **if-else** statements in the actions in the **switch** statement, only calls to the four functions.

```
if num < -2 || num > 4
        f1(num)
else
        if num < = 2
            if num > = 0
                f2(num)
        else
                f3(num)
            end
        else
            f4(num)
        end
end
```

24. In a certain city, there are five different trolley lines; they have the codes A, B, C, D, and E. Assume that *code* is a char variable that has been initialized. The following nested **if-else** statement prints the frequency with which a given line is supposed to run:

```
if (code > 'A' && code < 'E')
        if (code == 'C')
            fprintf('Every 10 minutes\n')
        else
            fprintf('Every 7 minutes\n')
        end
else
        if (code == 'A')
            fprintf('This line is now obsolete!\n')
```

```
        else
            if (code ~= 'E')
                fprintf('Error - there is no such trolley!\n')
            else
                fprintf('Every 12 minutes\n')
            end
        end
    end
```

This is confusing! Rewrite this as a **switch** statement that does *exactly* what the above nested **if-else** statement does, for any given value of the variable *code*. Do not use any **if** or **if-else** statements (there should only be **fprintf** statements in the **switch**). Put this in a script that will test the solution.

25. Write a script *area_menu* that will print a list consisting of cylinder, circle, and rectangle. It prompts the user to choose one, and then prompts the user for the appropriate quantities (e.g., the radius of the circle) and then prints its area. If the user enters an invalid choice, the script simply prints an error message. The script should use a nested **if-else** statement to accomplish this. Here are two examples of running it (units are assumed to be inches).

```
>> area_menu
Menu
1. Cylinder
2. Circle
3. Rectangle
Please choose one: 2
Enter the radius of the circle: 4.1
The area is 52.81

>> area_menu
Menu
1. Cylinder
2. Circle
3. Rectangle
Please choose one: 3
Enter the length: 4
Enter the width: 6
The area is 24.00
```

26. Modify the *area_menu* function to use a **switch** statement to decide which area to calculate.

27. Modify the *area_menu* function (either version) to use the built-in **menu** function instead of printing the menu choices.

28. Write a script that prompts the user for a value of a variable *x*. Then, it uses the **menu** function to present choices between sin(x), cos(x), and tan(x). The script will print whichever function of x the user chooses. Use an **if-else** statement to accomplish this.

29. Modify the preceding script to use a **switch** statement instead.
30. Write a function that will receive one number as an input argument. It will use the **menu** function that will display "Choose a function" and will have buttons labeled ceil, round, and sign. Using a **switch** statement, the function will then calculate and return the requested function (e.g., if round is chosen, the function will return the rounded value of the input argument).
31. Modify the previous function to use a nested **if-else** statement instead.

Looping

KEY WORDS

looping statement
counted loop
conditional loop
action
iterate
loop or iterator variable
running sum

running product
factorial
preallocate
echo-printing
nested loop
outer loop
inner loop

vectorizing
array operator
logical vector
infinite loop
counting
error-checking

Consider the problem of calculating the area of a circle with a radius of 0.3 centimeters—a MATLAB® program certainly is not needed to do that; you'd use your calculator instead, and punch in $\pi * 0.3^2$. However, if a table of circle areas is desired, for radii ranging from 0.1 centimeters to 100 centimeters in steps of 0.05 (e.g., 0.1, 0.15, 0.2, etc.), it would be very tedious to use a calculator and write it all down. This is one of the great uses of a computer—the ability to repeat a process such as this.

This chapter will cover statements in MATLAB that allow other statement(s) to be repeated. The statements that do this are called *looping statements*, or *loops*. There are two basic kinds of loops in programming: *counted loops*, and *conditional loops*. A counted loop is one that repeats statements a specified number of times (e.g., ahead of time it is known how many times the statements are to be repeated). In a counted loop, for example, you might say "repeat these statements 10 times." A conditional loop also repeats statements, but ahead of time it is not known *how many* times the statements will need to be repeated. With a conditional loop, for example, you might say "repeat these statements until this condition becomes false." The statement(s) that are repeated in any loop are called the action of the loop.

There are two different loop statements in MATLAB: the **for** statement and the **while** statement. In practice, the **for** statement usually is used as the counted loop, and the **while** is used as the conditional loop. To keep it simple, that is how they will be presented here.

Finally, although loops are extremely useful in most programming applications, in MATLAB they are frequently not necessary, especially when dealing with arrays. The concept of *vectorizing* will be introduced, which is the term used in MATLAB for rewriting code that was written using loops in a traditional programming language to matrix operations in MATLAB.

4.1 THE FOR LOOP

The **for** statement, or the **for** loop, is used when it is necessary to repeat statement(s) in a script or function, and when it *is* known ahead of time how many times the statements will be repeated. The statements that are repeated are called the action of the loop. For example, it may be known that the action of the loop will be repeated five times. The terminology used is that we *iterate* through the action of the loop five times.

The variable that is used to iterate through values is called a *loop variable*, or an *iterator variable*. For example, the variable might iterate through the integers 1 through 5 (e.g., 1, 2, 3, 4, and then 5). Although variable names in general should be mnemonic, it is common for an iterator variable to be given the name i (and if more than one iterator variable is needed, i, j, k, l, etc.) This is historical, and is because of the way integer variables were named in Fortran. However, in MATLAB both **i** and **j** are built-in values for $\sqrt{-1}$, so using either as a loop variable will override that value. If that is not an issue, then it is acceptable to use i as a loop variable.

The general form of the **for** loop is:

```
for loopvar = range
    action
end
```

where *loopvar* is the loop variable, *range* is the range of values through which the loop variable is to iterate, and the action of the loop consists of all statements up to the **end**. The range can be specified using any vector, but normally the easiest way to specify the range of values is to use the colon operator.

As an example, to print a column of numbers from 1 to 5:

```
for i = 1:5
    fprintf('%d\n',i)
end
```

This loop could be entered in the Command Window, although like **if** and **switch** statements, loops will make more sense in scripts and functions. In the Command Window, the results would appear after the **for** loop:

```
>> for i = 1:5
    fprintf('%d\n',i)
end
1
2
3
4
5
```

What the **for** statement accomplished was to print the value of i and then the newline character for every value of i, from 1 through 5 in steps of 1. The first thing that happens is that i is initialized to have the value 1. Then, the action of the loop is executed, which is the **fprintf** statement that prints the value of i (1), and then the newline character to move the cursor down. Then, i is incremented to have the value of 2. Next, the action of the loop is executed, which prints 2 and the newline. Then, i is incremented to 3 and that is printed, then i is incremented to 4 and that is printed, and then finally i is incremented to 5 and that is printed. The final value of i is 5; this value can be used once the loop has finished.

QUICK QUESTION!

How could you print this column of integers?

```
0
50
100
150
200
```

Answer: In a loop, you could print these values starting at 0, incrementing by 50 and ending at 200. Each is printed using a field width of 3.

```
>> for i = 0:50:200
    fprintf('%3d\n',i)
end
```

4.1.1 Finding Sums and Products

A very common application of a **for** loop is to calculate sums and products. For example, instead of just printing the integers 1 through 5, we could calculate the sum of the integers 1 through 5 (or, in general, 1 through *n*, where *n* is any positive integer). Basically, we want to implement

$$\sum_{i=1}^{n} i$$

or calculate the sum $1 + 2 + 3 + \ldots + n$.

In order to do this, we need to add each value to a ***running sum***. A running sum is a sum that will keep changing; we keep adding to it. First the sum has to be initialized to 0, then in this case it will be 1 $(0 + 1)$, then 3 $(0 + 1 + 2)$, then 6 $(0 + 1 + 2 + 3)$, and so forth.

In a function to calculate the sum, we need a loop or iterator variable *i*, as before, and also a variable to store the running sum. In this case we will use the output argument *runsum* as the running sum. Every time through the loop, the next value of *i* is added to the value of *runsum*. This function will return the end result, which is the sum of all integers from 1 to the input argument *n* stored in the output argument *runsum*.

sum_1_to_n.m

```
function runsum = sum_1_to_n(n)
% This function returns the sum of
% integers from 1 to n
runsum = 0;
for i = 1:n
   runsum = runsum + i;
end
```

As an example, if 5 is passed to be the value of the input argument *n*, the function will calculate and return $1 + 2 + 3 + 4 + 5$, or 15:

```
>> sum_1_to_n(5)
ans =
    15
```

Note that the output was suppressed when initializing the sum to 0 and when adding to it during the loop.

Another very common application of a **for** loop is to find a ***running product***. For example, instead of finding the sum of the integers 1 through n, we could find the product of the integers 1 through n. Basically, we want to implement

 PRACTICE 4.1

Write a function *sum_m_to_n* that is similar to the previous function but will calculate the sum of the integers from m to n. For example, if the integers 4 and 7 are passed to the function, it will calculate $4 + 5 + 6 + 7$:

```
>> sum_m_to_n(4,7)
ans =
    22
```

$$\prod_{i=1}^{n} i$$

or calculate the product 1 * 2 * 3 * 4 *... * n, which is called the *factorial* of n, written n!.

The Programming Concept

In order to do this, the basic algorithm is similar to finding a sum, except we need to multiply each value of the loop variable to a running product. The difference is that while a running sum variable is initialized to 0, a running product variable must be initialized to 1. This is so that the first time a value is multiplied by it, it does not change the original value.

```
myfact.m
function runprod = myfact (n)
% This function returns the product of
% integers from 1 to n, or n!
runprod = 1;
for i = 1:n
   runprod = runprod * i;
end
```

Any positive integer argument could be passed to this function, and it will calculate the factorial of that number. For example, if 5 is passed, the function will calculate and return 1 * 2 * 3 * 4 * 5, or 120:

```
>> myfact(5)
ans =
   120
```

The Efficient Method

MATLAB has a built-in function, **factorial**, that will find the factorial of an integer n:

```
>> factorial(5)
ans =
   120
```

4.1.1.1 Sums and Products with Vectors

The previous examples found either the sum or product of values at regular intervals, for example from 1 to an integer n. Frequently, however, we wish to find the sum and/or product of the elements in a vector, regardless of what those values might be. For example, we will write a function to sum all the elements in a vector.

The Programming Concept

The vector is passed as an argument to the function. The function loops through all the elements of the vector, from 1 to the length of the vector, to add them all to the running sum.

myvecsum.m

```
function outarg = myvecsum(vec)
% This function sums the elements in a vector
outarg = 0;
for i = 1:length(vec)
  outarg = outarg + vec(i);
end
```

Any vector could be passed to this function; for example, just specifying values for the elements in square brackets:

```
>> myvecsum([5 9 4])
ans =
     18
```

The Efficient Method

MATLAB has a built-in function, **sum**, that will sum all values in a vector. Again, any vector can be passed to the **sum** function:

```
>> sum([5 9 4])
ans =
     18
```

The function *myvecsum* illustrates a very important concept: looping through all the elements in a vector in order to do something with each one. In this case, we are adding every element in the vector to a running sum, which is stored in the output argument *outarg*. Notice that the loop variable, *i*, is used as the index into the vector. The first time through the loop, when *i* has the value 1, the value of vec(1), which is 5, is added to the value of *outarg* (so it is 0 + 5, or 5). The second time through the loop, when *i* has the value 2, vec(2) or 9 is added to *outarg* (so it now stores 0 + 5 + 9, or 14). Then, the third and final time through the loop vec(3) or 4 is added, so *outarg* now stores 0 + 5 + 9 + 4, or 18.

This is in fact one reason to store values in a vector. Values in a vector typically represent "the same thing", so in a program typically the same operation would be performed on every element. The general form of a **for** loop to accomplish this is:

```
for i = 1:length(vectorvariable)
   do something with vectorvariable(i)
end
```

The loop variable iterates through all elements in the vector, from 1 through the end (given by length(vectorvariable)), doing something with each element, specified as vectorvariable(i).

As another example, we will write a function to find the product of all the elements in a vector.

The Programming Concept

The vector is passed as an argument to the function. The function loops through all the elements of the vector, from 1 to the length of the vector, to multiply them all by the running product.

```
myvecprod.m
```

```
function outarg = myvecprod(vec)
% This function finds the product of
% the elements in a vector
outarg = 1;
for i = 1:length(vec)
   outarg = outarg * vec(i);
end
```

```
>> myvecprod(1:5)
ans =
   120
>> myvecprod([5 9 4])
ans =
   180
```

The Efficient Method

MATLAB has a built-in function, **prod**, that will return the product of all values in a vector.

```
>> prod([5 9 4])
   ans =
      180
```

QUICK QUESTION!

How could we write a function *prod_m_to_n* to calculate and return the product of the integers m to n without assuming a specific order of the arguments? In other words, both the function calls *prod_m_to_n(3,6)* and *prod_m_to_n(6,3)* would return the result of 3 * 4 * 5 * 6 or 360.

Answer: The Programming Concept

To loop from the smaller value to the larger, we would first have to compare their values and exchange them if necessary.

prod_m_to_n.m

```
function runprod = prod_m_to_n(m,n)
% Product of m to n using a for loop

% Make sure m is less than n
if m > n
  temp = m;
  m = n;
  n = temp;
end
% Loop to calculate the running product
runprod = 1;
for i = m:n
  runprod = runprod * i;
end
```

The Efficient Method

Instead of exchanging the values of *m* and **n**, we could use the colon operator with steps of either +1 or −1 to create a vector.

Also, instead of looping to calculate a running product, we could use the **prod** function.

prod_m_to_nii.m

```
function outprod = prod_m_to_nii(m,n)
% Product of m to n using : and prod
if m < n
  outprod = prod(m:n);
else
  outprod = prod(m:-1:n);
end
```

Additionally, MATLAB has functions **cumsum** and **cumprod** that return a vector of the running sums or products. For example, for the following vector the first value is 5, so that is the first value in the vector returned by **cumsum**. Then the next value is 9 so 5 + 9 is 14, and finally 14 + 4 is 18.

```
>> vec = [5 9 4];
>> cumsum(vec)
ans =
     5  14  18
```

The cumulative product function **cumprod** instead calculates 5, then 5 * 9, and finally 5 * 9 * 4:

```
>> cumprod(vec)
ans =
     5  45  180
```

4.1.1.2 Preallocating a Vector

There are essentially two programming methods that could be used to simulate the **cumsum** function. One method is to start with an empty vector and concatenate each running sum value to the vector. Extending a vector, however, is very inefficient. A better method is to *preallocate* the vector to the correct size and then change the value of each element to be successive running sums. Both methods will be shown here.

In the following function, the output argument is initialized to the empty vector []. Then, every time the next element in the vector is added to the running sum, this new sum is appended to the vector.

```
myveccumsum.m
```

```
function outvec = myveccumsum(vec)
% This function imitates cumsum for a vector
outvec = [];
runsum = 0;
for i = 1:length(vec)
  runsum = runsum + vec(i);
  outvec = [outvec runsum];
end
```

Here is an example of calling the function:

```
>> myveccumsum([5 9 4])
ans =
     5  14  18
```

The first time in the loop, *outvec* will be [5]. Then the second time, *runsum* will be 14 and outvec will store [5 14]. Finally, *runsum* will be 18 and *outvec* will store [5 14 18].

Although this method works, it is inefficient. Another method involves referring to each index in the output vector, and placing each partial sum into the next element in the output vector. As each value of *vec(i)* is added to the running sum, this new sum is stored in *outvec(i)*.

myveccumsumii.m

```
function outvec = myveccumsumii(vec)
% This function imitates cumsum for a vector
% It preallocates the output vector
outvec = zeros(size(vec));
runsum = 0;
for i = 1:length(vec)
  runsum = runsum + vec(i);
  outvec(i) = runsum;
end
```

Although initializing the output vector *outvec* to all zeros is not strictly necessary, it greatly improves the efficiency of the function. Initializing this vector to all zeros with the same size as the input argument preallocates that much memory for *outvec*. Then, each element is changed in the loop to its correct value. On the other hand, when building a vector up one element at a time, as in the previous example, MATLAB has to find new memory each time the vector is extended.

MATLAB has many other functions that work with vectors. Many, which are statistical in nature, will be seen in Chapter 12.

4.1.2 Combining for Loops with if Statements

 PRACTICE 4.2

Write a function that imitates the **cumprod** function. Use the method of preallocating the output vector.

Another example of a common application on a vector is to find the minimum or maximum value in the vector.

The Programming Concept

For example, to find the minimum value in a vector, the algorithm is:

- The minimum so far is the first element in the vector
- Loop through the rest of the vector
  - If any element is less than the minimum found so far, then that element is the new minimum so far

The following function implements this algorithm, and returns the minimum value found in the vector.

myminvec.m

```
function outmin = myminvec(vec)
% Finds the minimum value in a vector
outmin = vec(1);
for i = 2:length(vec)
  if vec(i) < outmin
    outmin = vec(i);
  end
end
```

```
>> vec = [3 8 99 -1];
>> myminvec(vec)
ans =
      -1
>> vec = [3 8 99 11];
>> myminvec(vec)
ans =
     3
```

Note the use of an **if** statement in the loop rather than an **if-else** statement. If the value of the next element in the vector is less than *outmin*, then the value of *outmin* is changed; otherwise, no action is necessary.

The Efficient Method

MATLAB has functions **min** and **max**, which find the minimum and maximum values in a vector.

```
>> vec = [5 9 4];
>> min(vec)
ans =

     4
```

4.1.3 For Loops that Do Not Use the Iterator Variable in the Action

In all the examples that we have seen so far, the value of the loop variable has been used in some way in the action of the **for** loop: we have printed the value of *i*, or added it to a sum, or multiplied it by a running product, or used it as an index into a vector. It is not always necessary to

 PRACTICE 4.3

Write a function to find and return the maximum value in a vector.

actually use the value of the loop variable, however. Sometimes the variable is simply used to iterate, or repeat, a statement a specified number of times. For example,

```
for i = 1:3
     fprintf('I will not chew gum\n')
end
```

produces the output:

```
I will not chew gum
I will not chew gum
I will not chew gum
```

The variable *i* is necessary to repeat the action three times, even though the value of *i* is not used in the action of the loop.

QUICK QUESTION!

What would be the result of the following **for** loop?

```
for i = 4:2:8
  fprintf('I will not chew gum\n')
end
```

Answer: Exactly the same output as before! It doesn't matter that the loop variable iterates through the values 4, then 6, then 8, instead of 1, 2, 3. Since the loop variable is not used in the action, this is just another way of specifying that the action should be repeated three times. Of course, using 1:3 makes more sense!

4.1.4 Input in a for Loop

The following script repeats the process of prompting the user for a number, and *echo printing* the number (which means simply printing it back out). A **for** loop specifies how many times this is to occur. This is another example in which the loop variable is not used in the action, but instead just specifies how many times to repeat the action.

forecho.m

```
% This script loops to repeat the action of
% prompting the user for a number and echo-printing it
for iv = 1:3
  inputnum = input('Enter a number: ');
  fprintf('You entered %.1f\n',inputnum)
end
```

```
>> forecho
Enter a number: 33
You entered 33.0
Enter a number: 1.1
You entered 1.1
Enter a number: 55
You entered 55.0
```

In this example, the loop variable *iv* iterates through the values 1 through 3, so the action is repeated three times. The action consists of prompting the user for a number and echo-printing it with one decimal place.

 PRACTICE 4.4

Modify the *forecho* script to sum the numbers that the user enters and print the result.

Instead of simply echo-printing the numbers, it is often necessary to store them in a vector. One way of accomplishing this is to start by preallocating the vector and then putting values in each element, as we saw in a previous example. Here is a function that does this, and returns the resulting vector. The function receives an input argument *n*, and repeats the process *n* times.

forinputvec.m

```
function numvec = forinputvec(n)
%Prompts the user and puts the numbers into a vector
numvec = zeros(1,n);
for iv = 1:n
  inputnum = input('Enter a number: ');
  numvec(iv) = inputnum;
end
```

Here is an example of calling this function and storing the resulting vector in a variable called *myvec*.

```
>> myvec = forinputvec(3)
Enter a number: 44
Enter a number: 2.3
Enter a number: 11
myvec =
   44.0000  2.3000  11.0000
```

QUICK QUESTION!

If you need to just print the sum or average of the numbers that the user enters, would you need to store them in a vector variable?

Answer: No. You could just add each to a running sum as you read them in a loop.

QUICK QUESTION!

What if you wanted to calculate how many of the numbers that the user entered were greater than the average?

Answer: Yes, then you would need to store them in a vector because you would have to go back through them to count how many were greater than the average (or, alternatively, you could go back and ask the user to enter them again!!).

4.2 NESTED FOR LOOPS

The action of a loop can be any valid statement(s). When the action of a loop is another loop, this is called a *nested loop*.

As an example, a nested **for** loop will be demonstrated in a script that will print a box of *'s. Variables in the script will specify how many rows and columns to print. For example, if *rows* has the value 3, and *columns* has the value 5, the output would be:

```
*****
*****
*****
```

Since lines of output are controlled by printing the newline character, the basic algorithm is:

- For every row of output,
  - Print the required number of *'s
  - Move the cursor down to the next line (print the '\n')

printstars.m

```
% Prints a box of stars
% How many will be specified by 2 variables
% for the number of rows and columns
rows = 3;
columns = 5;
% loop over the rows
for i=1:rows
  % for every row loop to print *'s and then one \n
  for j=1:columns
```

(Continued)

```
      fprintf('*')
   end
   fprintf('\n')
end
```

Running the script displays the output:

```
>> printstars
*****
*****
*****
```

The variable *rows* specifies the number of rows to print, and the variable *columns* specifies how many *'s to print in each row. There are two loop variables: *i* is the loop variable for the rows, and *j* is the loop variable for the columns. Since the number of rows and columns are known (given by the variables *rows* and *columns*), **for** loops are used. There is one **for** loop to loop over the rows, and another to print the required number of *'s. The values of the loop variables are not used within the loops, but are used simply to iterate the correct number of times. The first **for** loop specifies that the action will be repeated *rows* times. The action of this loop is to print *'s and then the newline character. Specifically, the action is to loop to print *columns* *'s across on one line. Then, the newline character is printed after all five stars to move the cursor down for the next line.

The first **for** loop is called the ***outer loop***; the second **for** loop is called the ***inner loop***. So, the outer loop is over the rows, and the inner loop is over the columns. The outer loop must be over the rows because the program is printing a certain number of rows of output. For each row, a loop is necessary to print the required number of *'s; this is the inner **for** loop.

When this script is executed, first the outer loop variable *i* is initialized to 1. Then, the action is executed. The action consists of the inner loop, and then printing the newline character. So, while the outer loop variable has the value 1, the inner loop variable *j* iterates through all its values. Since the value of *columns* is 5, the inner loop will print a * five times. Then, the newline character is printed and the outer loop variable *i* is incremented to 2. The action of the outer loop is then executed again, meaning the inner loop will print five *'s, and then the newline character will be printed. This continues, and in all, the action of the outer loop will be executed *rows* times.

Notice the action of the outer loop consists of two statements (the **for** loop and an **fprintf** statement). The action of the inner loop, however, is only a single statement.

The **fprintf** statement to print the newline character must be separate from the other **fprintf** statement that prints the *. If we simply had

```
fprintf('*\n')
```

as the action of the inner loop, this would print a long column of 15 *'s, not a box.

QUICK QUESTION!

How could this program be modified to print a triangle of *'s instead of a box? For example,

```
*
**
***
```

Answer: In this case, the number of *'s to print in each row is the same as the row number; for example, one * is printed in row 1, two *'s in row 2, and so on. So, it is not necessary to use a value for the number of columns; the inner **for** loop does not loop to *columns*, but to the value of the row loop variable:

printtristars.m

```
% Prints a triangle of stars
% How many will be specified by a variable
% for the number of rows
rows = 3:
for i=1:rows
    % inner loop just iterates to the value of i
    for j=1:i
        fprintf('*')
    end
    fprintf('\n')
end
```

```
>> printtristars
*
**
***
```

In these examples, the loop variables were used just to specify the number of times the action is to be repeated. These same loops could be used instead to produce a multiplication table by multiplying the values of the loop variables.

The following function *multtable* calculates and returns a matrix that is a multiplication table. Two arguments are passed to the function, which are the number of rows and columns for this matrix.

multtable.m

```
function outmat = multtable (rows, columns)
% Creates a matrix which is a multiplication table

% Preallocate the matrix
outmat = zeros(rows,columns);
for i = 1:rows
    for j = 1:columns
        outmat(i,j) = i * j;
    end
end
```

In the following example, the matrix has three rows and five columns:

```
>> multtable(3,5)
ans =
       1      2      3      4      5
       2      4      6      8     10
       3      6      9     12     15
```

Notice that this is a function that returns a matrix; it does not print anything. It preallocates the matrix to zeros, and then replaces each element. Since the number of rows and columns are known, **for** loops are used. The outer loop loops over the rows, and the inner loop loops over the columns. The action of the nested loop calculates i * j for all values of *i* and *j*. First, when *i* has the value 1, *j* iterates through the values 1 through 5, so first we are calculating 1 * 1, then 1 * 2, then 1 * 3, then 1 * 4, and finally 1 * 5. These are the values in the first row (first in element (1,1), then (1,2), then (1,3), then (1,4), and finally (1,5)). Then, when *i* has the value 2, the elements in the second row of the output matrix are calculated, as *j* again iterates through the values from 1 through 5. Finally, when *i* has the value 3, the values in the third row are calculated (3 * 1, 3 * 2, 3 * 3, 3 * 4, and 3 * 5).

This function could be used in a script that prompts the user for the number of rows and columns, calls this function to return a multiplication table, and writes the resulting matrix to a file:

createmulttab.m

```
% Prompt the user for rows and columns and
% create a multiplication table to store in
% a file mymulttable.dat
```

(Continued)

```
num_rows = input('Enter the number of rows: ');
num_cols = input('Enter the number of columns: ');
multmatrix = multtable(num_rows, num_cols);
save mymulttable.dat multmatrix -ascii
```

Here is an example of running this script, and then loading from the file into a matrix in order to verify that the file was created:

```
>> createmulttab
Enter the number of rows: 6
Enter the number of columns: 4
>> load mymulttable.dat
>> mymulttable
mymulttable =

    1     2     3     4
    2     4     6     8
    3     6     9    12
    4     8    12    16
    5    10    15    20
    6    12    18    24
```

4.2.1 Nested Loops and Matrices

Nested loops often are used when it is necessary to loop through all the elements of a matrix.

As an example, we will calculate the overall sum of the elements in a matrix.

The Programming Concept

The matrix is passed as an input argument to the function. The function then uses the **size** function to determine the number of rows and columns in the matrix. It then loops over all elements in the matrix by using a nested loop: one loop over the rows, and one loop over the columns, adding each element to the running sum. Notice that the loop variables i and j are used as the indices into the matrix: first the row index, and then the column index.

mymatsum.m

```
function outsum = mymatsum(mat)
% Calculates the overall sum of the elements
% in a matrix
[row col] = size(mat);
outsum = 0;
```

(Continued)

```
% The outer loop is over the rows
for i = 1:row
    for j = 1:col
        outsum = outsum + mat(i,j);
    end
end
```

```
>> mat = [3:5; 2 5 7]
mat =
     3    4    5
     2    5    7
>> mymatsum(mat)
ans =
    26
```

The Efficient Method

MATLAB has a built-in function **sum**, as we have seen. For matrices, like many built-in functions, the **sum** function operates columnwise, meaning that it will return the sum of each column.

```
>> mat
mat =
     3    4    5
     2    5    7
>> sum(mat)
ans =
     5    9    12
```

So, to get the overall sum, it is necessary to sum the column sums!

```
>> sum(sum(mat))
ans =
    26
```

In this example, the outer loop was over the rows, and the inner loop was over the columns. This order could easily be switched, however, so that the outer loop is over the columns and the inner loop is over the rows:

```
mymatsumb.m
```

```
function outsum = mymatsumb(mat)
% Calculates the overall sum of the elements
% in a matrix
```

(Continued)

```
[row col] = size(mat);
outsum = 0;
% The outer loop is over the columns
for i = 1:col
    for j = 1:row
        outsum = outsum + mat(j,i);
    end
end
```

```
>> mat = [3:5; 2 5 7]
mat =
        3    4    5
        2    5    7
>> mymatsumb(mat)
ans =
       26
```

The order of the loops does not matter in this example, since all that is required is to add each element to the overall sum. Notice that when referring to an element in the matrix, the row index is always given first, and then the column index, regardless of the order of the loops.

QUICK QUESTION!

How would we sum each individual column, rather than just getting an overall sum?

Answer: The programming method would require a nested loop in which the outer loop is over the columns. For example, we will modify the previous function to sum each column and return a row vector containing the results.

matcolsum.m

```
function outsum = matcolsum(mat)
% Calculates the sum of every column in a matrix
% Returns a vector of the column sums
[row col] = size(mat);
% Preallocate the vector to the number of columns
outsum = zeros(1,col);
% Every column is being summed so the outer loop
```

(Continued)

```
% has to be over the columns
for i = 1:col
      % Initialize the running sum to 0 for every column
      runsum = 0;
      for j = 1:row
            runsum = runsum + mat(j,i);
      end
      outsum(i) = runsum;
end
```

Notice that the output argument will be a row vector containing the same number of columns as the input argument matrix. Also, since the function is calculating a sum for each column, the *runsum* variable must be initialized to 0 for every column, so it is initialized inside of the outer loop.

```
>> mat = [3:5; 2 5 7]
mat =
        3    4    5
        2    5    7
```

```
>> matcolsum(mat)
ans =
        5    9    12
```
Of course, the built-in **sum** function in MATLAB would accomplish the same thing, as we have already seen.

Note that since the built-in **sum** function sums each column, one way of using **sum** to find the sum of each row is to transpose the matrix:

```
>> sum(mat')
ans =
        12    14
```

For matrices, the **cumsum** function returns a matrix consisting of the first row of the matrix argument, then the sum of the values in the first and second rows (for a matrix that has two rows).

```
>> cumsum(mat)
ans =
        3    4    5
        5    9    12
```

 PRACTICE 4.5

Modify this function; create a function *matrowsum* to calculate and return a vector of all the row sums, instead of column. For example, calling it and passing the previous *mat* variable would result in the following:

```
>> matrowsum(mat)
ans =
        12    14
```

The functions **min** and **max** also operate columnwise; these functions find the minimum or maximum values in each column.

```
>> mat
mat =
     3    4    5
     2    5    7
>> max(mat)
ans =
     3    5    7
```

4.2.2 Combining Nested for Loops and if Statements

The statements inside of a nested loop can be any valid statement, including any selection statement. For example, there could be an **if** or **if-else** statement as the action, or part of the action, in a loop.

As an example, there is a file called datavals.dat, which has results recorded from an experiment. However, some were erroneously recorded. The numbers are all supposed to be positive. The following script reads from this file into a matrix. It prints the sum from each row of only the positive numbers. We will assume that the file contains integers but will not assume how many lines are in the file nor how many numbers per line.

sumonlypos.m

```
% Sums only positive numbers from file
% Reads from the file into a matrix and then
% calculates and prints the sum of only the
% positive numbers from each row
load datavals.dat
[r c] = size(datavals);
for i = 1:r
    sumrow = 0;
    for j = 1:c
        if datavals(i,j) >=0
            sumrow = sumrow + datavals(i,j);
        end
    end
    fprintf('The sum for row %d is %d\n',i,sumrow)
end
```

For example, *if* the file contains:

33	−11	2
4	5	9
22	5	−7
2	11	3

the output from the program would look like this:

```
>> sumonlypos
The sum for row 1 is 35
The sum for row 2 is 18
The sum for row 3 is 27
The sum for row 4 is 16
```

The file is loaded into a matrix variable. The script finds the size of the matrix and then loops through all the elements in the matrix by using a nested loop; the outer loop iterates through the rows and the inner loop iterates through the columns. For each element, an **if-else** statement determines whether the element is positive or not. It only adds the positive values to the row sum. Since the sum is found for each row, the *sumrow* variable is initialized to 0 for every row, meaning inside the outer loop.

 PRACTICE 4.6

Write a function *mymatmin* that finds the minimum value in each column of a matrix argument and returns a vector of the column minimums. Here is an example of calling the function.

```
>> mat = randint(3,4,[1 20])
mat =
    15    19    17     5
     6    14    13    13
     9     5     3    13
>> mymatmin(mat)
ans =
     6  5  3  5
```

QUICK QUESTION!

Would it matter if the order of the loops was reversed in this example, so the outer loop iterates over the columns and the inner loop over the rows?

Answer: Yes, since we want a sum for every row the outer loop must be over the rows.

QUICK QUESTION!

Would the function *mymatmin* also work for a vector argument?

Answer: Yes, it should, since a vector is just a subset of a matrix. In this case, one of the loops would repeat only one time (for the rows if it is a row vector, or for the columns if it is a column vector).

4.3 VECTORIZING

In many cases in MATLAB, loops are not necessary. Since MATLAB is written specifically to work with vectors and matrices, many operations can be done and functions can be called with vectors and matrices. In fact, the term *vectorizing* is used in MATLAB for rewriting code using loops in a traditional programming language to matrix operations in MATLAB.

In most programming languages, when performing an operation on a vector, a **for** loop is used to loop through the entire vector. For example, in MATLAB assuming there is a vector variable *vec*:

```
for i = 1:length(vec)
    % do something with vec(i)
end
```

Similarly, for an operation on a matrix, a nested loop would be required; for example, assuming a matrix variable *mat*:

```
[r c] = size(mat);
for row = 1:r
    for col = 1:c
        % do something with mat(row,col)
    end
end
```

Usually in MATLAB, this is not necessary.

Numerical operations can be done on entire vectors or matrices. For example, let's say that we want to multiply every element of a vector v by 3, and store the result back in v, where v is initialized as follows:

```
>> v = [3 7 2 1];
```

The Programming Concept

To accomplish this, we can loop through all the elements in the vector and multiply each element by 3. In the following, the output is suppressed in the loop, and then the resulting vector is shown:

```
>> for i = 1:length(v)
    v(i) = v(i) * 3;
end
>> v
v =
    9  21  6  3
```

The Efficient Method

In MATLAB, we can simply multiply v by 3 and store the result back in v in an assignment statement:

```
>> v = v*3
v =
    9  21  6  3
```

As another example, we can divide every element by 2:

```
>> v= [3 7 2 1];
>> v/2
ans =
     1.5000   3.5000   1.0000   0.5000
```

For a matrix, numerical operations can also be performed on every element. For example, to multiply every element in a matrix by 2 with most languages would involve a nested loop, but in MATLAB it is automatic.

```
>> mat = [4:6; 3: -1:1]
mat =
     4     5     6
     3     2     1
>> mat * 2
ans =
     8    10    12
     6     4     2
```

However, for the exponentiation operator, .^ must be used when working with vectors and matrices, rather than just the ^. This is true actually when using any operator based on multiplication (which means multiplication, division, and exponentiation) and any vector or matrix term-by-term. Squaring a vector, for example, means multiplying each element by itself so the .^ operator must be used.

```
>> v ^ 2
??? Error using ==> mpower
Matrix must be square.

>> v .^ 2
ans =
     9    49     4     1
```

The operators .^, .*, ./, and .\ are called *array operators*.

4.3.1 Logical Vectors

The relational operators can also be used with vectors and matrices. For example, let's say that there is a vector, and we want to compare every element in the vector to 5 to determine whether it is greater than 5 or not. The result would be a vector (with the same length as the original) with logical true or false values. Assume a variable vec as shown here.

```
>> vec = [5 9 3 4 6 11];
```

 PRACTICE 4.7

- Create a vector variable and add 2 to every element in it.
- Create a matrix variable and divide every element by 3.
- Create a matrix variable and square every element.

The Programming Concept

To accomplish this using the programming method, we would have to loop through all the elements of the vector and compare each element with 5 to determine whether the corresponding value in the result would be logical true or false.

The Efficient Method

In MATLAB, this can be accomplished automatically by simply using the relational operator >.

```
>> isg = vec > 5
isg =
     0  1  0  0  1  1
```

Notice that this creates a vector consisting of all logical true or false values. Although this is a vector of ones and zeros, and numerical operations can be done on the vector *isg*, its type is **logical** rather than **double**.

```
>> doubres = isg + 5
ans =
     5  6  5  5  6  6

>> whos
```

Name	Size	Bytes	Class
doubres	1x6	48	double array
isg	1x6	6	logical array
vec	1x6	48	double array

To determine how many of the elements in the vector *vec* were greater than 5, the **sum** function could be used on the resulting vector *isg*:

```
>> sum(isg)
ans =
     3
```

The *logical vector isg* can also be used to index into the vector. For example, if only the elements from the vector that are greater than 5 are desired:

```
>> vec(isg)
ans =
     9  6  11
```

QUICK QUESTION!

Why doesn't this work?

```
>> vec([0 1 0 0 1 1])
??? Subscript indices must either be
real positive integers or logicals.
```

Answer: The difference between the vector in this example and *isg* is that *isg* is a vector of logicals (**logical** 1's and 0's), whereas [0 1 0 0 1 1] by default is a vector of **double** values. Only **logical** 1's and 0's can be used to index into a vector.

Because the values in the vector must be **logical** 1's and 0's, the following function that appears at first to accomplish the same operation using the programming method, actually does not. The function receives two input arguments: the vector, and an integer with which to compare (so it is somewhat more general). It loops through every element in the input vector, and stores in the result vector either a 1 or 0 depending on whether vec(i) > n is true or false.

testvecgtn.m

```
function outvec = testvecgtn(vec,n)
% Compare each element in vec to see whether it
% is greater than n or not

% Preallocate the vector
outvec = zeros(size(vec));
for i = 1:length(vec)
  % Each element in the output vector stores 1 or 0
  if vec(i) > n
    outvec(i) = 1;
  else
    outvec(i) = 0;
  end
end
```

Calling the function appears to return the same vector as simply vec > 5, and summing the result still works to determine how many elements were greater than 5.

```
>> notlog = testvecgtn(vec,5)
notlog =
     0  1  0  0  1  1
```

```
>> sum(notlog)
ans =
     3
```

However, as before, it could not be used to index into a vector because the elements are **double**, not **logical**:

```
>> vec(notlog)
??? Subscript indices must either be real positive integers
or logicals.
```

QUICK QUESTION!

How could we remedy this? How can we assign logical values rather than doubles?

Answer: We can use the **logical** function to make the output argument vector the type **logical** rather than **double**. In the function, the only statement that needs to be modified is

the statement that preallocates the vector; in this case, it also changes the type of the vector to **logical**.

```
outvec = logical(zeros(size(vec)));
```

4.3.1.1 Logical Built-In Functions

There are built-in functions in MATLAB that are useful in conjunction with vectors or matrices of all logical true or false values; two of these are the functions **any** and **all**. The function **any** returns logical true if any element in a vector is logically true, and false if not. The function **all** returns logical true only if all elements are logically true. Here are some examples. For the variable *vec1*, all elements are logical true so both **any** and **all** return true.

```
>> vec1 = [1 3 1 1 2];
>> any(vec1)
ans =
     1
>> all(vec1)
ans =
     1
```

For *vec2*, some elements are logical true so **any** returns true but **all** returns false.

```
>> vec2 = [1 1 0 1]
vec2 =
     1     1     0     1
>> any(vec2)
ans =
     1
```

```
>> all(vec2)
ans =
     0
```

QUICK QUESTION!

How could we accomplish the same thing by using loops and **if** statements?

Answer: For the equivalent to the **any** function, we would start by assuming that the result is logical false; that is, that none of the elements are logical true, and then loop through the elements, setting the result instead to logical true if anything in the vector is found to be logically true.

myany.m

```
function logresult = myany(vec)
% Simulates the built-in function any

% Assume 0 for the result
logresult = logical(0);
for i = 1:length(vec)
  % if any value is not false, the result will be 1
  if vec(i) ~= 0
    logresult = logical(1);
  end
end
```

```
>> vec2 = [1 1 0 1];
>> myany(vec2)
ans =
     1
>> vec3 = [0 0 0];
>> myany(vec3)
ans =
     0
```

This works, but it is inefficient in that it checks every element in the vector even if a value has already been found that is logically true. (We will see how to fix this in the next section using a conditional loop instead of the **for** loop.)

For the equivalent to the **all** function, we must make sure that all the elements in the vector are logically true. One way of doing this is to count how many elements are true, and then compare the count to the number of elements in the vector. Recall that any nonzero value can be used to represent the concept of true.

myall.m

```
function logresult = myall(vec)
% Simulates the built-in function all

% count how many values are true
count = 0;
for i = 1:length(vec)
  if vec(i) ~= 0
    count = count + 1;
  end
end
% if all were true, return 1 else return 0
if count == length(vec)
  logresult = logical(1);
else
  logresult = logical(0);
end
```

```
>> myall(vec1)
ans =
     1
>> myall(vec2)
ans =
     0
>> myall(vec3)
ans =
     0
```

The function **find** returns the indices of a vector that meet some criteria. For example, to find all the elements in a vector that are greater than 5:

```
>> vec = [5 3 6 7 2]
vec =
  5  3  6  7  2
>> find(vec > 5)
ans =
       3    4
```

Also, the function **isequal** is useful in comparing vectors. In MATLAB, using the equality operator with arrays will return 1 or 0 for each element; the **all** function could then be used on the resulting array to determine whether all elements were equal or not. The built-in function **isequal** also accomplishes this:

```
>> vec1 = [1 3 -4 2 99];
>> vec2 = [1 2 -4 3 99];
>> vec1 == vec2
ans =
     1     0     1     0     1
>> all(vec1 == vec2)
ans =
     0
>> isequal(vec1,vec2)
ans =
     0
```

MATLAB also has **or** and **and** operators that work elementwise for matrices:

Operator	Meaning
|	elementwise or for matrices
&	elementwise and for matrices

These operators will compare any two vectors or matrices, as long as they are the same size, element-by-element, and return a vector or matrix of the same size of logical 1's and 0's. The operators || and && are used only with scalars, not matrices. For example,

```
>> v1 = [3 0 5 1];
>> v2 = [0 0 2 0];
>> v1 & v2
ans =
     0     0     1     0
>> v1 | v2
ans =
     1     0     1     1
>> v1 && v2
??? Operands to the || and && operators must be convertible
to logical scalar values.
```

Table 4.1 Operator Precedence Rules

Operators	Precedence
parentheses ()	highest
transpose and power ', ^, .^	
unary negation (–), not (~)	
multiplication, division *,/,\,.*,./,.\	
addition, subtraction +, –	
colon operator :	
relational <, <=, >, >=, ==, ~=	
element-wise and &	
element-wise or |	
and &&	
or ||	
assignment =	lowest

As with the numerical operators, it is important to know the operator precedence rules. Table 4.1 shows the rules for the operators that have been covered so far, in the order of precedence.

4.3.2 Vectors and Matrices as Function Arguments

Using most programming languages, if it is desired to evaluate a function on every element in a vector or a matrix, a loop would be necessary to accomplish this. However, as we have already seen, in MATLAB an entire vector or matrix can be passed as an argument to a function; the function will be evaluated on every element. This means that the result will be the same size as the argument.

The Programming Method

For example, let us find the sine in radians of every element of a vector *vec*. The algorithm would be to loop through the elements of the vector and get the sine of each one. To store the results in a new vector, the most efficient way would be to preallocate it.

```
>> vec = −2:1
vec =
        −2  −1  0  1
>> sinvec = zeros(size(vec));
```

```
>> for i = 1:length(vec)
     sinvec(i) = sin(vec(i));
   end
>> sinvec
sinvec =
    -0.9093  -0.8415  0  0.8415
```

The Efficient Method

The **sin** function will automatically return the sine of each individual element and the result will also be a vector with a length of four (in this case, in *ans*).

```
>> sin(vec)
ans =
    -0.9093 -0.8415 0 0.8415
```

For a matrix, the resulting matrix will have the same size as the input argument matrix. For example, the **sign** function will find the sign of each element in a matrix:

```
>> mat = [0 4 -3; -1 0 2]
mat =
    0   4  -3
   -1   0   2
>> sign(mat)
ans =
    0   1  -1
   -1   0   1
```

The Programming Method

To write our own *signum* function that accomplishes exactly the same thing as the built-in **sign** function, nested loops would be required.

signum.m

```
function outmat = signum(mat)
% This function imitates the sign function
[r c] = size(mat);
for i = 1:r
  for j = 1:c
    if mat(i,j) > 0
      outmat(i,j) = 1;
    elseif mat(i,j) == 0
```

(Continued)

```
        outmat(i,j) = 0;
    else
        outmat(i,j) = -1;
    end
  end
end
```

To test this function, we will create a matrix of random integers in a range from –8 to 8, and pass this to the function. Here are some examples of using this function:

```
>> mat = randint(2,4,[-8 8])
mat =
    -2     7    -3     2
     0     2     0    -6
>> signum(mat)
ans =
    -1     1    -1     1
     0 ·   1     0    -1
>> signmat = signum(mat)
signmat =
    -1     1    -1     1
     0     1     0    -1

>> help signum
    This function imitates the sign function
```

Vectors or matrices can be passed to user-defined functions as well, as long as the operators used in the function are correct. For example, we previously defined a function that calculates the area of a circle:

```
>> type calcarea
   function area = calcarea(rad)
   % This function calculates the area of a circle
   area = pi * rad * rad;
```

This function was written assuming that the argument was a scalar, so calling it with a vector instead would produce an error message:

```
>> calcarea(1:3)
??? Error using ==> mtimes
Inner matrix dimensions must agree.
Error in ==> calcarea at 3
   area = pi * rad * rad;
```

This is because the * was used for multiplication in the function, but .* must be used when multiplying vectors term-by-term. Changing this in the function would allow either scalars or vectors to be passed to this function:

calcareaii.m

```
function area = calcareaii(rad)
% This function calculates the area of a circle
% The input argument can be a vector of radii
area = pi * rad .* rad;
```

```
>> calcareaii(1:3)
ans =
   3.1416  12.5664  28.2743
>> calcareaii(4)
ans =
   50.2655
```

4.4 WHILE LOOPS

The **while** statement is used as the conditional loop in MATLAB; it is used to repeat an action when ahead of time it is *not* known *how many* times the action will be repeated. The general form of the **while** statement is:

```
while condition
      action
end
```

The action, which consists of any number of statement(s), is executed as long as the condition is true. The condition must eventually become false to avoid an *infinite loop*. (If this happens, Ctrl-C will exit the loop.)

The way it works is that first the condition is evaluated. If it is logically true, the action is executed. So, to begin with it is just like an **if** statement. However, at that point the condition is evaluated again. If it is still true, the action is executed again. Then, the action is evaluated again. If it is still true, the action is executed again. Then, the action is… eventually, this has to stop! Eventually something in the action has to change something in the condition so it becomes false.

As an example of a conditional loop, we will write a function that will find the first factorial that is greater than the input argument *high*. Previously, we wrote a function to calculate a particular factorial. For example, to calculate 5! we found the product 1 * 2 * 3 * 4 * 5. In that case a **for** loop was used, since it was known that the loop would be repeated five times. Now, we do not know how many times the loop will be repeated. The basic algorithm is to have two variables, one that iterates through the values 1, 2, 3, and so on, and one that stores the factorial of the iterator at each step. We start with 1, and 1 factorial, which is 1. Then, we check the factorial. If it is not greater than *high*, the iterator variable will then increment to 2, and find its factorial (2). If this is not greater than *high*, the iterator will then increment to 3, and the function will find its factorial (6). This continues until we get to the first factorial that is greater than *high*. So, the process of incrementing a variable and finding its factorial is repeated until we get to the first value greater than *high*. This is implemented using a **while** loop:

factgthigh.m

```
function facgt = factgthigh(high)
% Finds the first factorial > high
i=0;
fac=1;
while fac <= high
   i=i+1;
   fac = fac * i;
end
facgt = fac;
```

Here is an example of calling the function, passing 5000 for the value of the input argument *high*.

```
>> factgthigh(5000)
ans =
    5040
```

The iterator variable *i* is initialized to 0, and the running product variable *fac*, which will store the factorial of each value of *i*, is initialized to 1. The first time the **while** loop is executed, the condition is conceptually true: 1 is less than or equal to 5000. So, the action of the loop is executed, which is to increment *i* to 1 and *fac* to 1 (1 * 1). After the execution of the action of the loop, the condition is evaluated again. Since it will still be true, the action is executed: *i* is incremented to 2, and *fac* will get the value 2 (1 * 2). The value 2 is still <= 5000, so the action will be executed again: *i* will be incremented to 3, and *fac* will get the value 6 (2 * 3). This continues until the first value of *fac* is found that is greater than 5000. As soon as *fac* gets to this value, the condition will be false and the **while** loop will end. At that point the factorial is assigned to the output argument, which returns the value.

The reason that *i* is initialized to 0 rather than 1 is that the first time the loop action is executed, *i* becomes 1 and *fac* becomes 1 so we have 1 and 1!, which is 1.

Notice that the output of all assignment statements is suppressed in the function.

4.4.1 **Multiple Conditions in a** While **Loop**

In the previous section, we wrote a function *myany* that imitated the built-in **any** function by returning logical true if any value in the input vector was logical true, and logical false otherwise. The function was inefficient because it looped through all the elements in the input vector, even though once one logical true value is found it is no longer necessary to examine any other elements. A **while** loop will improve on this. Instead of looping through all the elements, what we really want to do is to loop until either a logical true value is found, or until we've gone through the entire vector. Thus, we have two parts to the condition in the **while** loop. In the following function, we initialize the output argument to logical false, and an iterator variable i to 1. The action of the loop is to examine an element from the input vector: if it is logical true, we change the output argument to be logical true. Also in the action the iterator variable is incremented. The action of the loop is continued as long as the index has not yet reached the end of the vector, and as long as the output argument is still logical false.

myanywhile.m

```
function logresult = myanywhile(vec)
% Simulates the built-in function any
% Uses a while loop so that the action halts
% as soon as any true value is found
logresult = logical(0);
i = 1;
while i <= length(vec) && logresult == 0
  if vec(i) ~= 0
    logresult = logical(1);
  end
  i = i + 1;
end
```

The output produced by this function is the same as the *myany* function, but it is more efficient because now as soon as the output argument is set to logical true, the loop ends.

4.4.2 **Reading from a File in a While Loop**

The previous example demonstrated using an **if** statement in a **while** loop. Although in many languages the combination of a loop and an **if** statement would be necessary to determine whether or not the elements in a vector are logical true, MATLAB has built-in functions such as **any**, **all**, and **find** to accomplish these tasks.

For example, data from an experiment has been recorded in a file called 'experd. dat'. The file has some numbers followed by a −99 and then more numbers, all on the same line. The only data values that we are interested in, however, are those before the −99. The algorithm for the script will be:

1. Read the data from the file into a vector.

2. Create a new vector variable *newvec* that has the data values only up to but not including the −99.

3. Plot the new vector values, using black o's.

For example, if the file has the following:

```
3.1 11 5.2 8.9 −99 4.4 62
```

FIGURE 4.1

Plot of some (but not all) data from a file.

the plot produced would look like Figure 4.1.

For simplicity, we will assume that the file format is as specified. Using **load** will create a vector with the name *experd*, which contains the values from the file. Also, since this is generic data we will omit the plot labels and titles for simplicity.

The Programming Concept

Using the programming method, we would loop through the vector until the −99 is found, creating the new vector by storing each element from *experd* in the vector *newvec*.

findvalwhile.m

```
% Reads data from a file, but only plots the numbers
% up to a flag of −99. Uses a while loop.
load experd.dat
i = 1;
while experd(i) ~= −99
  newvec(i) = experd(i);
  i = i + 1;
end
plot(newvec,'ko')
```

The Efficient Method

Using the **find** function, we can locate the index of the element that stores the −99. Then, the new vector is all the original vector from the first element to the index before the index of the element that stores the −99.

findval.m

```
% Reads data from a file, but only plots the numbers
% up to a flag of -99. Uses find and the colon operator.
load experd.dat
where = find(experd == -99);
newvec = experd(1:where-1);
plot(newvec,'ko')
```

4.4.3 **Input in a** While **Loop**

The following script repeats the process of prompting the user, reading in a positive number, and echo-printing it, as long as the user correctly enters positive numbers when prompted. As soon as the user types in a negative number, the program will print OK and end.

whileposnum.m

```
% Prompts the user and echo prints the numbers entered
% until the user enters a negative number
inputnum=input('Enter a positive number: ');
while inputnum >= 0
   fprintf('You entered a %d.\n\n',inputnum)
   inputnum = input('Enter a positive number: ');
end
fprintf('OK!\n')
```

When the program is executed, the input/output might look like this:

```
>> whileposnum
Enter a positive number: 6
You entered a 6.

Enter a positive number: -2
OK!
```

If the user enters a negative number the first time, no values would be echo-printed:

```
>> whileposnum
Enter a positive number: -33
OK!
```

This illustrates a very important feature of **while** loops: it is possible that the action will not be executed at all, if the value of the condition is false the first time it is evaluated.

As we have seen before, MATLAB will give an error message if a character is entered rather than a number.

```
>> whileposnum
Enter a positive number: a
??? Error using ==> input
Undefined function or variable 'a'.

Enter a positive number: -4
OK!
```

However, if the character is actually the name of a variable, it will use the value of that variable as the input. For example:

```
>> a = 5;
>> whileposnum
Enter a positive number: a
You entered a 5.

Enter a positive number: -4
OK!
```

4.4.4 Counting in a While Loop

When it is not known ahead of time how many values will be entered into a script, it is frequently necessary to *count* the number of values that are entered. For example, if numbers are read into a script, and then the average of the numbers is desired, the script must add them together, and keep track of how many there are, in order to calculate the average. The following variation on the previous script counts the number of numbers that the user successfully enters:

countposnum.m

```
% Prompts the user for positive numbers and echo prints as
% long as the user enters positive numbers

% Counts the positive numbers entered by the user
counter=0;
inputnum=input('Enter a positive number: ');
while inputnum >= 0
   fprintf('You entered a %d.\n\n',inputnum)
   counter = counter + 1;
   inputnum = input('Enter a positive number: ');
end
fprintf('Thanks, you entered %d positive numbers\n',counter)
```

The script initializes a variable *counter* to 0. Then, in the **while** loop action, every time the user successfully enters a number, the program increments the counter variable. At the end of the script, it prints the number of numbers that were entered.

```
>> countposnum
Enter a positive number: 4
You entered a 4.

Enter a positive number: 8
You entered a 8.

Enter a positive number: 11
You entered a 11.

Enter a positive number: -4
Thanks, you entered 3 positive numbers
```

4.4.5 Error-Checking User Input in a While **Loop**

In most applications, when the user is prompted to enter something, there is a valid range of values. If the user enters an incorrect value, rather than having the program carry on with an incorrect value, or just printing an error message, the program should repeat the prompt. The program should keep prompting the user, reading the value, and checking it, until the user enters a value that is in the correct range. This is a very common application of a conditional loop, looping until the user correctly enters a value in a program. This is called *error-checking*.

 PRACTICE 4.8

Write a script *aveposnum* that will repeat the process of prompting the user for positive numbers, until the user enters a negative number, as earlier. Instead of echo-printing them, however, the script will print the average (of just the positive numbers). If no positive numbers are entered, the script will print an error message instead of the average. Here are examples of executing this script:

```
>> aveposnum
Enter a positive number: -5
No positive numbers to
average.

>> aveposnum
Enter a positive number: 8
Enter a positive number: 3
Enter a positive number: 4
Enter a positive number: -6
The average was 5.00
```

For example, the following script prompts the user to enter a positive number, and loops to print an error message and repeat the prompt until the user finally enters a positive number.

readonenum.m

```
% Loop until the user enters a positive number
inputnum=input('Enter a positive number: ');
while inputnum < 0
   inputnum = input('Invalid! Enter a positive number: ');
end
fprintf('Thanks, you entered a %.1f \n',inputnum)
```

Here is an example of running this script:

```
>> readonenum
Enter a positive number: -5
Invalid! Enter a positive number: -2.2
Invalid! Enter a positive number: c
??? Error using ==> input
Undefined function or variable 'c'.

Invalid! Enter a positive number: 44
Thanks, you entered a 44.0
```

Note that MATLAB itself catches the character input and prints an error message and repeats the prompt when the c was entered.

QUICK QUESTION!

How could we vary this example, so that the script asks the user to enter positive numbers *n* times, where *n* is an integer defined to be 3?

Answer: Every time the user enters a value, the script checks and in a **while** loop keeps telling the user that it's invalid until a valid positive number is read. By putting the error-check in a **for** loop that repeats n times, the user is forced eventually to enter three positive numbers.

readnnums.m

```
% Loop until the user enters n positive numbers
n=3;
fprintf('Please enter %d positive numbers\n\n',n)
for i=1:n
    inputnum=input('Enter a positive number: ');
    while inputnum < 0
        inputnum = input('Invalid! Enter a positive number: ');
    end
    fprintf('Thanks, you entered a %.1f \n',inputnum)
end
```

```
>> readnnums
Please enter 3 positive numbers

Enter a positive number: 5.2
Thanks, you entered a 5.2
```

```
Enter a positive number: 6
Thanks, you entered a 6.0
Enter a positive number: -7.7
Invalid! Enter a positive number: 5
Thanks, you entered a 5.0
```

4.4.5.1 Error-Checking for Integers

Since MATLAB uses the type **double** by default for all values, to check to make sure that the user has entered an integer, the program has to convert the input value to an integer type (e.g., **int32**) and then check to see whether that is equal to the original input. The following examples illustrate the concept.

If the value of the variable *num* is a real number, converting it to the type **int32** will round it, so the result is not the same as the original value.

```
>> num = 3.3;
>> inum = int32(num)
inum =
        3
>> num == inum
ans =
      0
```

If, on the other hand, the value of the variable *num* is an integer, converting it to an integer type will not change the value.

```
>> num = 4;
>> inum = int32(num)
inum =
          4
>> num == inum
ans =
     1
```

The following script uses this idea to error-check for integer data; it loops until the user correctly enters an integer.

readoneint.m

```
% Error-check until the user enters an integer
inputnum = input('Enter an integer: ');
num2 = int32(inputnum);
while num2 ~= inputnum
   inputnum = input('Invalid! Enter an integer: ');
   num2 = int32(inputnum);
end
fprintf('Thanks, you entered a %d \n',inputnum)
```

Here are examples of running this script:

```
>> readoneint
Enter an integer: 9.5
```

```
Invalid! Enter an integer: 3.6
Invalid! Enter an integer: -11
Thanks, you entered a -11

>> readoneint
Enter an integer: 5
Thanks, you entered a 5
```

Putting these ideas together, the following script loops until the user correctly enters a positive integer. There are two parts to the condition, since the value must be positive and must be an integer.

readoneposint.m

```
% Error checks until the user enters a positive integer
inputnum = input('Enter a positive integer: ');
num2 = int32(inputnum);
while num2 ~= inputnum || num2 < 0
  inputnum = input('Invalid! Enter a positive integer: ');
  num2 = int32(inputnum);
end
fprintf('Thanks, you entered a %d \n',inputnum)
```

 PRACTICE 4.9

Modify this script to read *n* positive integers, instead of just one.

```
>> readoneposint
Enter a positive integer: 5.5
Invalid! Enter a positive integer: -4
Invalid! Enter a positive integer: 11
Thanks, you entered a 11
```

SUMMARY

Common Pitfalls

- Forgetting to initialize a running sum or count variable to 0.

- Forgetting to initialize a running product variable to 1.

- In cases where loops are necessary, not realizing that if an action is required for every row in a matrix, the outer loop must be over the rows (and if an action is required for every column, the outer loop must be over the columns).

- Attempting to use an array of **double** 1's and 0's to index into an array (must be **logical**, instead).

- Forgetting that for array operations based on multiplication, the dot must be used in the operator. In other words, for multiplying, dividing, or raising to an exponent term-by-term, the operators are .*, ./, and .^.

- Attempting to use || or && with arrays; always use | and & when working with arrays; || and && are used only with scalars.

- Not realizing that it is possible that the action of a **while** loop will never be executed.

- Not error-checking input into a program.

Programming Style Guidelines

- Use loops for repetition only when necessary:
  - **for** statements as counted loops
  - **while** statements as conditional loops

- Do not use *i* or *j* for iterator variable names if the use of the built-in constants **i** and **j** is desired.

- Indent the action of a loop.

- If the loop variable is being used just to specify how many times the action of the loop is to be executed, use the colon operator 1:*n* where *n* is the number of times the action is to be executed.

- Preallocate vectors and matrices whenever possible (when the size is known ahead of time).

- When data is read in a loop, store it in an array only if it will be necessary to access the individual data values again.

- Vectorize code whenever possible. If it is not necessary to use loops in MATLAB, don't!

- Use the array operators .*, ./, .\, and .^ in functions so that the input arguments can be arrays and not just scalars.

MATLAB Functions and Commands			
while	sum	cumprod	any
for	prod	min	all
end	cumsum	max	find
factorial			

MATLAB Operators
element-wise or for matrices |
element-wise and for matrices &

Exercises

1. Write a **for** loop that will print the column of real numbers from 1.1 to 2.9 in steps of 0.1.

2. Write a function **sumsteps2** that calculates and returns the sum of 1 to n in steps of 2, where n is an argument passed to the function. For example, if 11 is passed, it will return $1 + 3 + 5 + 7 + 9 + 11$. Do this using a **for** loop. Calling the function will look like this:

   ```
   >> sumsteps2(11)
   ans =
      36
   ```

3. Write a function **prodby2** that will receive a value of a positive integer n and will calculate and return the product of the odd integers from 1 to n (or from 1 to $n–1$ if n is even).

4. Write a function called **geomser** that will receive values of r and n, and will calculate and return the sum of the geometric series:

   $$1 + r + r^2 + r^3 + r^4 + \ldots + r^n$$

   The following examples of calls to this function illustrate what the result should be:

   ```
   >> geomser(1,5)
   ans =
         6
   >> disp(geomser(2,4))
         31
   ```

5. Prompt the user for an integer n and print "I love this stuff!" n times.

6. In the Command Window, write a **for** loop that will iterate through the integers from 32 to 255. For each, show the corresponding character from the character encoding.

7. In the Command Window, write a **for** loop that will print the elements from a vector variable in sentence format. For example, if this is the vector:

   ```
   >> vec = [5.5 11 3.45];
   ```

   this would be the result:

   ```
   Element 1 is 5.50.
   Element 2 is 11.00.
   Element 3 is 3.45.
   ```

The **for** loop should work regardless of how many elements are in the vector.

8. Create a 1 × 6 vector of random integers, each in the range from 1 to 20. Use built-in functions to find the minimum and maximum values in the vector. Also create a vector of cumulative sums using **cumsum**.

9. Write a relational expression for a vector variable that will verify that the last value in a vector created by **cumsum** is the same as the result returned by **sum**.

10. Type this script and be amazed by the results! Try more points to get a clearer picture, but don't go overboard—it might take a while on your computer.

```
clear
clf
x = rand;
y = rand;
plot(x,y)
hold on
for it = 1:2000
  choic = round(rand*2);
  if choic == 0
    x = x/2;
    y = y/2;
  elseif choic == 1
    x = (x+1)/2;
    y = y/2;
  else
    x = (x+0.5)/2;
    y = (y+1)/2;
  end
  plot(x,y)
  hold on
end
```

11. A sound engineer has recorded a sound signal from a microphone. The sound signal was sampled, meaning that values at discrete intervals were recorded (rather than a continuous sound signal). The units of each data sample are volts. The microphone was not on at all times, however, so that data samples below a certain threshold are considered to be data values that were samples when the microphone was not on, and therefore not valid data samples. The sound engineer would like to know the average voltage of the sound signal. Write a script that will ask the user for the threshold and the number of data samples, and then for the individual data samples. The program will then print the average and a count of the valid data samples, or an error message if there were no valid data samples. An example of what the input and output would look like in the Command Window is shown:

```
Please enter the threshold below which samples will be
considered to be invalid:
```

```
3.0
Please enter the number of data samples to be entered:
7

Please enter a data sample: 0.4
Please enter a data sample: 5.5
Please enter a data sample: 5.0
Please enter a data sample: 2.1
Please enter a data sample: 6.2
Please enter a data sample: 0.3
Please enter a data sample: 5.4

The average of the 4 valid data samples is 5.53 volts.
```

Note: If there had been no valid data samples, the program would print an error message instead of the last line shown.

12. Write a script that will load data from a file into a matrix. Create the data file first, and make sure that there is the same number of values on every line in the file so that it can be loaded into a matrix. Using a **for** loop, it will then create as many Figure Windows as there are rows in the matrix, and will plot the numbers from each row in a separate Figure Window. For example, if the file contains:

```
4            9            22
30           18           4
```

two Figure Windows would be created, as seen in Figure 4.2.

13. A machine cuts N pieces of a pipe. After each cut, each piece of pipe is weighed and its length is measured; these two values are then stored in a file called 'pipe. dat' (first the weight and then the length on each line of the file). Ignoring units, the weight is supposed to be between 2.1 and 2.3, inclusive, and the length is supposed to be between 10.3 and 10.4, inclusive. The following is just the

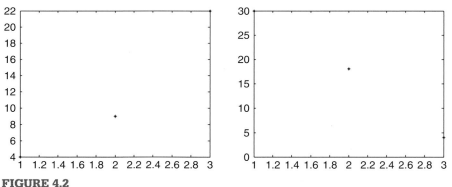

FIGURE 4.2

Number of plots from data file specified in a **for** *loop.*

beginning of what will be a long script to work with this data. For now, the script will just count how many rejects there are. A reject is any piece of pipe that has an invalid weight and/or length. For a simple example, if N is 3 (meaning three lines in the file) and the file stores:

```
2.14 10.30
2.32 10.36
2.20 10.35
```

there is only one reject, the second one, which weighs too much. The script would print:

```
There were 1 rejects.
```

14. Write a function that will receive a matrix as an input argument, and will calculate and return the overall average of all numbers in the matrix. Use loops, not built-in functions, to calculate the average.

15. We have seen that by default, when using built-in functions like **sum**, **prod**, and such on matrices, MATLAB will perform the function on each column. A dimension can also be specified when calling these functions; MATLAB refers to the columns as dimension 1 and the rows as dimension 2, for example,

```
>> sum(mat,1)
>> sum(mat,2)
```

Create a matrix and find the product of each row and column using **prod**.

16. Create a vector of five random integers, each in the range from −10 to 10. Perform each of the following two ways: using built-in functions, and also using loops (with **if** statements if necessary):

- Subtract 3 from each element.
- Count how many are positive.
- Get the absolute value of each element.
- Find the maximum.

17. Create a 3 × 5 matrix. Perform each of the following two ways: using built-in functions, and also using loops (with **if** statements if necessary):

- Find the maximum value in each column.
- Find the maximum value in each row.
- Find the maximum value in the entire matrix.

18. Write a script that will print the following multiplication table:

```
1
2  4
3  6  9
4  8 12 16
5 10 15 20 25
```

19. The Wind Chill Factor (WCF) measures how cold it feels with a given air temperature T (in degrees Fahrenheit) and wind speed (V, in miles per hour). One formula for it is

    $$WCF = 35.7 + 0.6T - 35.7\,(V^{0.16}) + 0.43T\,(V^{0.16})$$

    Write a function to receive the temperature and wind speed as input arguments, and return the WCF. Using loops, print a table showing wind chill factors for temperatures ranging from –20 to 55 in steps of 5, and wind speeds ranging from 0 to 55 in steps of 5. Call the function to calculate each wind speed.

20. Instead of printing the wind chill factors in the previous problem, create a matrix of wind chill factors and write them to a file.

21. Generate a random integer n, create a vector of the integers 1 through n in steps of 2, square them, and plot the squares.

22. A vector v stores for several employees of the Green Fuel Cells Corporation their hours worked one week followed for each by the hourly pay rate. For example, if the variable stores

    ```
    >> v
    v =
        33.0000 10.5000 40.0000 18.0000 20.0000 7.5000
    ```

    that means the first employee worked 33 hours at $10.50 per hour, the second worked 40 hours at $18 an hour, and so on. Write code that will separate this into two vectors, one that stores the hours worked and another that stores the hourly rates. Then, use the array multiplication operator to create a vector, storing in the new vector the total pay for every employee.

23. The mathematician Euler proved the following:

    $$\frac{\pi^2}{6} = 1 + \frac{1}{4} + \frac{1}{9} + \frac{1}{16} + \ldots$$

    Rather than finding a mathematical proof for this, try to verify whether the conjecture seems to be true or not. Note: There are two basic ways to approach this: either choose a number of terms to add, or loop until the sum is close to $\pi^2/6$.

24. Write a script that will prompt the user for the radius and height of a cone, error-check the user's input for the radius and the height, and then calculate and print the volume of the cone.

25. Write a script (for example, called *findmine*) that will prompt the user for minimum and maximum integers, and then another integer that is the user's choice in the range from the minimum to the maximum. The script will then generate random integers in the range from the minimum to the maximum, until a match for the user's choice is generated. The script will print how many random integers had to be generated until a match for the user's choice was found. For example, running this script might result in this output:

```
>> findmine
Please enter your minimum value: -2
Please enter your maximum value: 3
Now enter your choice in this range: 0
It took 3 tries to generate your number
```

26. Write a script that will prompt the user for N integers, and then write the positive numbers (>= 0) to an ASCII file called pos.dat and the negative numbers to an ASCII file called neg.dat. Error-check to make sure that the user enters N integers.

27. Write a script that will continue prompting the user for positive numbers, and storing them in a vector variable, until the user types a negative number.

28. Write a script that will use the **menu** function to present the user with choices for functions **fix**, **floor**, and **ceil**. Error-check by looping to display the menu until the user pushes one of the buttons (an error could occur if the user clicks the X on the menu box rather than pushing one of the buttons). Then, generate a random number and print the result of the user's function choice of that number (e.g., **fix(5)**).

29. Write a script *echoletters* that will prompt the user for letters of the alphabet and echo-print them until the user enters a character that is not a letter of the alphabet. At that point, the script will print the nonletter, and a count of how many letters were entered. Here are examples of running this script:

```
>> echoletters
Enter a letter: T
Thanks, you entered a T
Enter a letter: a
Thanks, you entered a a
Enter a letter: 8
8 is not a letter
You entered 2 letters

>> echoletters
Enter a letter: !
! is not a letter
You entered 0 letters
```

The format must be exactly as shown.

30. Write a script called *prtemps* that will prompt the user for a maximum Celsius value in the range from -16 to 20; error-check to make sure it's in that range. Then, print a table showing degrees F and degrees C until this maximum is reached. The first value that exceeds the maximum should not be printed. The table should start at 0 degrees F, and increment by 5 degrees F until the max (in C) is reached. Both temperatures should be printed with a field width of 6 and one decimal place. The formula is C = 5/9 (F – 32). For example, the execution of the script might look like this (the format should be exactly like this):

```
>> prtemps
When prompted, enter a temp in degrees C in range -16
to 20.
Enter a maximum temp: 30
Error! Enter a maximum temp: 9
     F              C
   0.0          -17.8
   5.0          -15.0
  10.0          -12.2
  15.0           -9.4
  20.0           -6.7
  25.0           -3.9
  30.0           -1.1
  35.0            1.7
  40.0            4.4
  45.0            7.2
```

31. Create an x vector that has integers 1 through 10, and set a y vector equal to x. Plot this straight line. Now, add noise to the data points by creating a new y^2 vector that stores the values of $y \pm 0.25$. Plot the straight line and also these noisy points.

32. A blizzard is a massive snowstorm. Definitions vary, but for our purposes we will assume that a blizzard is characterized by both winds of 30 mph or higher and blowing snow that leads to visibility of a half-mile or less, sustained for at least four hours. Data from a storm one day has been stored in a file stormtrack.dat. There are 24 lines in the file, one for each hour of the day. Each line in the file has the wind speed and visibility at a location. Create a sample data file. Read this data from the file and determine whether blizzard conditions were met during this day or not.

MATLAB Programs

Chapter 2 introduced scripts and user-defined functions. In that chapter, we saw how to write script files, which are sequences of statements that are stored in an M-file and then executed. We also saw how to write user-defined functions, also stored in M-files, that calculate and return a single value. In this chapter, we will expand on these concepts, and introduce other kinds of user-defined functions. We will show how MATLAB® programs consist of combinations of scripts and user-defined functions. The mechanisms for interactions of variables in M-files and the Command Window will be explored. Finally, techniques for finding and fixing mistakes in programs will be reviewed.

5.1 MORE TYPES OF USER-DEFINED FUNCTIONS

We already have seen how to write a user-defined function, stored in an M-file, that calculates and returns one value. This is just one type of function. It is also possible for a function to return multiple values, and it is possible for a function to return nothing. We will categorize functions as follows:

- Functions that calculate and return one value

- Functions that calculate and return more than one value

- Functions that just accomplish a task, such as printing, without returning any values

So, although many functions calculate and return values, some do not. Some functions instead just accomplish a task. Regardless of what kind of function it is, all functions must be defined, and all function definitions consist of the header and the body. Also, the function must be called in order for it to be utilized. Categorizing the functions is somewhat arbitrary, but there are differences between these three types of functions, including the format of the function headers and also the way in which the functions are called.

In general, any function in MATLAB consists of

- The function *header* (the first line); this has:

  - the reserved word **function**

  - (if the function *returns* values, the name(s) of the output argument(s) followed by the assignment operator =)

  - the name of the function (*Important*: This should be the same as the name of the M-file in which this function is stored in order to avoid confusion)

  - the input arguments in parentheses, if there are any (separated by commas if there is more than one)

- A comment that describes what the function does (this is printed if **help** is used)

- The *body* of the function, which includes all statements, including assigning values to all output arguments if there are any

5.1.1 Functions that Return More than One Value

Functions that return one value have one output argument, as we have seen in Section 2.7. Functions that return more than one value must instead have more than one output argument in the function header, in square brackets. That means that in the body of the function, values must be assigned to all output arguments listed in the function header. The general form of a function definition for a function that calculates and ***returns more than one value*** looks like this:

functionname.m

```
function [output arguments] = functionname(input arguments)
% Comment describing the function
Statements here; these must include assigning values to
all of the output arguments listed in the header
```

In the vector of output arguments, the output argument names by convention are separated by commas.

For example, here is a function that calculates two values, both the area and the circumference of a circle; this is stored in a file called 'areacirc.m':

areacirc.m

```
function [area, circum] = areacirc(rad)
% This function calculates the area and
% the circumference of a circle
area = pi * rad .* rad;
circum = 2 * pi * rad;
```

Since this function is calculating two values, there are two output arguments in the function header (*area* and *circum*), which are placed in square brackets [ ]. Therefore, somewhere in the body of the function, values have to be assigned to both. Since the function is returning two values, it is important to capture and store them in separate variables when the function is called, just as both the number of rows and the number of columns would be captured from the **size** function. In this case, the first value returned, which is the area of the circle, is stored in a variable *a* and the second value returned is stored in a variable *c*.

```
>> [a c] = areacirc(4)
a =
    50.2655
c =
    25.1327
```

If this is not done, only the first value returned is retained; in this case, the area:

```
>> disp(areacirc(4))
    50.2655
```

Note that in capturing the values the order matters. In this case, the function returns first the area and then the circumference of the circle. The order in which values are assigned to the output arguments within the function, however, does not matter.

QUICK QUESTION!

What would happen if a vector of radii was passed to the function?

Answer: Since the .* operator is used in the function to square *rad*, a vector can be passed to the input argument *rad*. Therefore, the results will also be vectors, so the variables on the left-hand side of the assignment operator would become vectors of areas and circumferences.

```
>> [a c] = areacirc(1:4)
a =
    3.1416 12.5664 28.2743 50.2655
c =
    6.2832 12.5664 18.8496 25.1327
```

The **help** function shows the comment listed under the function header:

```
>> help areacirc
  This function calculates the area and
  the circumference of a circle
```

The *areacirc* function could be called from the Command Window as shown here, or from a script. Here is a script that will prompt the user for the radius of just one circle, call the *areacirc* function, to calculate and return the area and circumference of the circle, and print the results:

calcareacirc.m

```
% This script prompts the user for the radius of a circle,
% calls a function to calculate and return both the area
% and the circumference, and prints the results
% It ignores units and error-checking for simplicity
radius = input('Please enter the radius of the circle: ');
[area circ] = areacirc(radius);
fprintf('For a circle with a radius of %.1f,\n', radius)
fprintf(' the area is %.1f and the circumference is %.1f\n',...
area, circ)
```

```
>> calcareacirc
Please enter the radius of the circle: 5.2
For a circle with a radius of 5.2,
    the area is 84.9 and the circumference is 32.7
```

 PRACTICE 5.1

Write a function **perimarea** that calculates and returns the perimeter and area of a rectangle. Pass the length and width of the rectangle as input arguments. For example, this function might be called from this script:

calcareaperim.m

```
% Prompt the user for the length and width of a rectangle,
% call a function to calculate and return the perimeter
% and area, and print the result
% For simplicity it ignores units and error-checking
length = input('Please enter the length of the rectangle: ');
width = input('Please enter the width of the rectangle: ');
[perim area] = perimarea(length, width);
fprintf('For a rectangle with a length of %.1f and a', length);
fprintf(' width of %.1f,\nthe perimeter is %.1f,', width, perim);
fprintf(' and the area is %.1f\n', area)
```

As another example, consider a function that calculates and returns three output arguments. The function will receive one input argument representing a total number of seconds, and returns the number of hours, minutes, and remaining seconds that it represents. For example, 7515 total seconds is 2 hours, 5 minutes, and 15 seconds because 7515 = 3600 * 2 + 60 * 5 + 15.

The algorithm is:

- Divide the total seconds by 3600, which is the number of seconds in an hour. For example, 7515/3600 is 2.0875. The integer part is the number of hours, for example, 2.

- The remainder of the total seconds divided by 3600 is the remaining number of seconds; it is useful to store this in a local variable.

■ The number of minutes is the remaining number of seconds divided by 60 (again, the integer part).

■ The number of seconds is the remainder of this division.

breaktime.m

```
function [hours, minutes, secs] = breaktime(totseconds)
% This function breaks a total number of seconds into
% hours, minutes, and remaining seconds
hours = floor(totseconds/3600);
remsecs = rem(totseconds, 3600);
minutes = floor(remsecs/60);
secs = rem(remsecs,60);
```

Here is an example of calling this function:

```
>> [h m s] = breaktime(7515)
h =
    2
m =
    5
s =
    15
```

As before, it is important to store all values that the function returns in variables.

5.1.2 Functions that Accomplish a Task Without Returning Values

Many functions do not calculate values, but rather accomplish a task such as printing formatted output. Since these functions do not return any values, there are no output arguments in the function header.

The general form of a function definition for a *function that does not return any values* looks like this:

functionname.m

```
function functionname(input arguments)
% Comment describing the function
Statements here
```

Notice what is missing in the function header: there are no output arguments, and no assignment operator.

For example, the following function just prints the number arguments passed to it in a sentence format:

```
printem.m
```
```
function printem(a,b)
% This function prints two numbers in a sentence format
fprintf('The first number is %.1f and the second is %.1f\n',a,b)
```

Since this function isn't calculating anything, there aren't any output arguments in the function header, and no =. An example of a call to the function is:

```
>> printem(3.3, 2)
The first number is 3.3 and the second is 2.0
```

Note that since the function isn't returning any value, it cannot be called from an assignment statement. Any attempt to do this would result in an error, for example,

```
>> x = printem(3, 5) % Error!!
??? Error using ==> printem
Too many output arguments.
```

We can therefore think of the call to a function that does not return values as a statement by itself, in that the function call cannot be imbedded in another statement such as an assignment statement or an output statement.

5.1.3 Functions that Return Values Versus Printing

A function that calculates and *returns* values (through the output arguments) does not normally also print them; that is left to the calling script or function. It is good programming practice to separate these tasks.

If a function just prints a value, rather than returning it, the value cannot be used later on in other calculations. For example, here is a function that just prints the circumference of a circle:

 PRACTICE 5.2

Write a function that receives a vector as an input argument and prints the elements from the vector in a sentence format.

```
>> printvecelems([5.9 33 11])
Element 1 is 5.9
Element 2 is 33.0
Element 3 is 11.0
```

```
calccircum1.m
```
```
function calccircum1(radius)
% displays the circumference of a circle but does not
% return the value
disp(2 * pi * radius)
```

Calling this function prints the circumference, but there is no way to store the value so that it can be used in subsequent calculations:

```
>> calccircum1(3.3)
   20.7345
```

Since no value is returned by the function, it would be an error to try to store the value in a variable:

```
>> c = calccircum1(3.3)
??? Error using ==> calccircum1
Too many output arguments.
```

By contrast, the following function calculates and returns the circumference, so that it can be stored and used in other calculations. For example, if the circle is the base of a cylinder, and we wish to calculate the surface area of the cylinder, we would need to multiply the result from the *calccircum2* function by the height of the cylinder.

calccircum2.m

```
function circle_circum = calccircum2(radius)
% calculates and returns the circumference of a circle
circle_circum = 2 * pi * radius;
```

```
>> circumference = calccircum2(3.3)
circumference =
   20.7345

>> height = 4;
>> surf_area = circumference * height
surf_area =
   82.9380
```

5.1.4 Passing Arguments to Functions

In all these function examples so far, at least one argument was passed in the function call to be the value(s) of the corresponding input argument(s) in the function header. The method used has been the ***call-by-value*** method of passing the values of the arguments to the input arguments in the functions.

In some cases, however, it is not necessary to pass any arguments to the function. Consider, for example, a function that simply prints a random real number with two decimal places:

printrand.m

```
function printrand()
% This function prints one random number
fprintf('The random # is %.2f\n',rand)
```

Here is an example of calling this function:

```
>> printrand()
The random # is 0.94
```

Since nothing is passed to the function, there are no arguments in the parentheses in the function call, and none in the function header, either. In fact, the parentheses are not even needed in either the function or the function call; this works, also:

printrandnp.m

```
function printrandnp
% This function prints one random number
fprintf('The random # is %.2f\n',rand)
```

```
>> printrandnp
The random # is 0.52
```

This was an example of a function that did not receive any input arguments nor did it return any output arguments; it simply accomplished a task.

Here is another example of a function that does not receive any input arguments, but in this case it does return a value. The function prompts the user for a string and returns the value entered.

stringprompt.m

```
function outstr = stringprompt
% Prompt the user for a string and return it
disp('When prompted, enter a string of any length.')
outstr = input('Enter the string here: ', 's');
```

```
>> mystring = stringprompt
When prompted, enter a string of any length.
Enter the string here: Hi there
mystring =
Hi there
```

 PRACTICE 5.3

Write a function that will prompt the user for a positive number, loop to error-check to make sure that the number is positive, and return the positive number.

QUICK QUESTION!

It is important that the number of arguments in the call to a function be the same as the number of input arguments in the function header, even if that number is zero. Also, if a function returns more than one value, it is important to capture all values by having an equivalent number of variables in a vector on the left-hand side of an assignment statement. Although it is not an error if there aren't enough variables, the values returned will be lost. The following question is posed to highlight this.

Given the following function header (note: this is just the function header, not the entire function definition):

```
function [outa, outb] = qq1(x, y, z)
```

which of the following proposed calls to this function would be valid?

```
a) [var1 var2] = qq1(a, b, c);
b) answer = qq1(3, y, q);
c) [a b] = myfun(x, y, z);
d) [outa outb] = qq1(x, z);
```

Answer: The first proposed function call, (a), is valid. There are three arguments that are passed to the three input arguments in the function header, the name of the function is *qq1*, and there are two variables in the assignment statement to store the two values returned from the function. Function call (b) is valid, although only the first value returned from the function would be stored in *answer*, the second value would be lost. Function call (c) is invalid because the name of the function is given incorrectly. Function call (d) is invalid because only two arguments are passed to the function, but there are three input arguments in the function header.

5.2 MATLAB PROGRAM ORGANIZATION

A MATLAB program typically consists of a script that calls functions to do the actual work.

5.2.1 Modular Programs

In a *modular program*, the solution is broken down into modules, and each is implemented as a function. The script is typically called the *main program*.

In order to demonstrate the concept, we will use the very simple example of calculating the area of a circle. In the next section, a much longer and more realistic example will be given. For this example, there are three steps in the algorithm to calculate the area of a circle:

- Get the input (the radius)

- Calculate the area

- Display the results

In a modular program, there would be one main script that calls three separate functions to accomplish these tasks:

- A function to prompt the user and read in the radius

- A function to calculate the area of the circle

- A function to display the results

Since both scripts and functions are stored in M-files, there would be four separate M-files altogether for this program; one M-file script and three M-file functions, as follows:

calcandprintarea.m

```
% This is the main script to calculate the
% area of a circle
% It calls 3 functions to accomplish this
radius = readradius;
area = calcarea(radius);
printarea(radius,area)
```

readradius.m

```
function radius = readradius
%This function prompts the user and reads the radius
disp('When prompted, please enter the radius in inches.')
radius = input('Enter the radius: ');
```

calcarea.m

```
function area = calcarea(rad)
% This function calculates the area of a circle
area = pi * rad * rad;
```

printarea.m

```
function printarea(rad,area)
% This function prints the results
fprintf('For a circle with a radius of %.2f inches,\n',rad)
fprintf('the area is %.2f inches squared.\n',area)
```

When the program is executed, the following steps will take place:

- The script *calcandprintarea* begins executing.

- *calcandprintarea* calls the *readradius* function:

    - *readradius* executes and returns the radius.

- *calcandprintarea* resumes executing and calls the *calcarea* function, passing the radius to it:

    - *calcarea* executes and returns the area.

- *calcandprintarea* resumes executing and calls the *printarea* function, passing both the radius and the area to it:
  - *printarea* executes and prints.

Running the program would be accomplished by typing the name of the script; this would call the other functions:

```
>> calcandprintarea
When prompted, please enter the radius in inches.
Enter the radius: 5.3
For a circle with a radius of 5.30 inches,
the area is 88.25 inches squared.
```

Notice how the function calls and the function headers match up. For example:

readradius function:
```
function call: radius = readradius;
function header: function radius = readradius
```

In the function call, no arguments are passed so there are no input arguments in the function header. The function returns one output argument so that is stored in one variable.

calcarea function:
```
function call: area = calcarea(radius);
function header: function area = calcarea(rad)
```

In the function call, one argument is passed in parentheses so there is one input argument in the function header. The function returns one output argument so that is stored in one variable.

printarea function:
```
function call: printarea(radius,area)
function header: function printarea(rad,area)
```

In the function call, there are two arguments passed, so there are two input arguments in the function header. The function does not return anything, so the call to the function is a statement by itself; it is not in an assignment or output statement.

Of course, the *readradius* function should really error-check the user's input.

 PRACTICE 5.4

Modify the *readradius* function to error-check the user's input to make sure that the radius is valid. The function should ensure that the radius is a positive number by looping to print an error message until the user enters a valid radius.

5.2.2 Subfunctions

So far, we have put each function in a separate M-file. However, it is possible to have more than one function in a given M-file. For example, if one function calls another, the first (calling) function would be the **primary function**, and the function that is called is a **subfunction**. These functions both would be stored in the same M-file—first the primary function and then the subfunction. The name of the M-file would be the same as the name of the primary function, to avoid confusion.

To demonstrate this, a program that is similar to the previous one, but calculates and prints the area of a rectangle, is shown here. The script, or main program, first calls a function that reads the length and width of the rectangle, and then calls a function to print the results. This function calls a subfunction to calculate the area.

rectarea.m

```
% This program calculates & prints the area of a rectangle

% Call a fn to prompt the user & read the length and width
[length, width] = readlenwid;
% Call a fn to calculate and print the area
printrectarea(length, width)
```

readlenwid.m

```
function [l,w] = readlenwid
% This function prompts the user for the length and width
l = input('Please enter the length: ');
w = input('Please enter the width: ');
```

printrectarea.m

```
function printrectarea(len, wid)
% This function prints the rectangle area
% It calls a subfunction to calculate the area
area = calcrectarea(len,wid);
fprintf('For a rectangle with a length of %.2f\n',len)
fprintf('and a width of %.2f, the area is %.2f\n', ...
    wid, area);

function area = calcrectarea(len, wid)
% This function calculates the rectangle area
area = len * wid;
```

This is an example of running this program:

```
>> rectarea
Please enter the length: 6
Please enter the width: 3
For a rectangle with a length of 6.00
and a width of 3.00, the area is 18.00
```

Notice how the function calls and function headers match up. For example:

readlenwid function:

```
function call: [length, width] = readlenwid;
function header: function [l,w] = readlenwid
```

In the function call, no arguments are passed so there are no input arguments in the function header. The function returns two output arguments so there is a vector with two variables on the left-hand side of the assignment statement in which the function is called.

printrectarea function:

```
function call: printrectarea(length, width)
function header: function printrectarea(len, wid)
```

In the function call, there are two arguments passed, so there are two input arguments in the function header. The function does not return anything, so the call to the function is a statement by itself; it is not in an assignment or output statement.

calcrectarea subfunction:

```
function call: area = calcrectarea(len,wid);
function header: function area = calcrectarea(len, wid)
```

In the function call, two arguments are passed in parentheses so there are two input arguments in the function header. The function returns one output argument so that is stored in one variable.

The **help** command can be used with the script *rectarea*, the function *readlenwid*, and the primary function *printrectarea*. To view the first comment in the subfunction, since it is contained within the printrectarea.m file, the operator > is used to specify both the primary and subfunctions:

```
>> help rectarea
   This program calculates & prints the area of a rectangle
>> help printrectarea
   This function prints the rectangle area
   It calls a subfunction to calculate the area
>> help printrectarea>calcrectarea
   This function calculates the rectangle area
```

5.3 APPLICATION: MENU-DRIVEN MODULAR PROGRAM

Many longer, more involved programs that have interaction with the user are **menu-driven**, which means that the program prints a menu of choices and then continues to loop to print the menu of choices until the user chooses to end

PRACTICE 5.5

For a right triangle with sides *a*, *b*, and *c*, where *c* is the hypotenuse and θ is the angle between sides *a* and *c*, the lengths of sides *a* and *b* are given by:

```
a = c cos(θ)
b = c sin(θ)
```

Write a script *righttri* that calls a function to prompt the user and read in values for the hypotenuse and the angle (in radians), and then calls a function to calculate and return the lengths of sides *a* and *b*, and a function to print out all values in a nice sentence format. For simplicity, ignore units. Here is an example of running the script; the output format should be exactly as shown here:

```
>> righttri
Enter the hypotenuse: 5
Enter the angle: .7854
For a right triangle with hypotenuse 5.0
  and an angle 0.79 between side a & the hypotenuse,
 side a is 3.54 and side b is 3.54
```

For extra practice, do this using two different program organizations:

- One script that calls three separate functions
- One script that calls two functions; the function that calculates the lengths of the sides will be a subfunction to the function that prints

the program. A modular menu-driven program typically would have a function that presents the menu and gets the user's choice, as well as functions to implement the action for each choice. These functions may have subfunctions. Also, the functions would error-check all user input.

As an example of such a menu-driven program, we will write a program to explore the constant *e*.

The constant *e*, called the natural exponential base, is used extensively in mathematics and engineering. There are many diverse applications of this constant. The value of the constant *e* is approximately 2.1718. Raising *e* to the power of x, or e^x, is so common that this is called the exponential function. In MATLAB, as we have seen, there is a function for this, **exp**.

One way to determine the value of *e* is by finding a limit.

$$e = \lim_{n \to \infty} \left(1 + \frac{1}{n}\right)^n$$

As the value of *n* increases toward infinity, the result of this expression approaches the value of *e*.

An approximation for the exponential function can be found using what is called a Maclaurin series:

$$e^x \approx 1 + \frac{x^1}{1!} + \frac{x^2}{2!} + \frac{x^3}{3!} + \dots$$

We will write a program to investigate the value of e and the exponential function. It will be menu-driven. The menu options will be:

- Print an explanation of e.
- Prompt the user for a value of n, and then find an approximate value for e using the expression $(1 + 1/n)^n$.
- Prompt the user for a value for x. Print the value of **exp(x)** using the built-in function. Find an approximate value for e^x using the Maclaurin series just given.
- Exit the program.

The algorithm for the script main program is:

- Call a function *eoption* to display the menu and return the user's choice.
- Loop until the user chooses to exit the program. If the user has not chosen to exit, the action of the loop, depending on the user's choice, either is to:
  - Call a function *explaine* to print an explanation of e
  - Call a function *limite* that will prompt the user for n and calculate an approximate value for e
  - Prompt the user for x and call a function *expfn* that will print both an approximate value for e^x and the value of the built-in **exp(x)**. (Note: Any value for x is acceptable so the program does not need to error-check this value.)
- Call the function *eoption* to display the menu and return the user's choice again.

The algorithm for the *eoption* function is:

- Use the **menu** function to display the four choices.
- Error-check (an error would occur if the user clicks the X on the menu box rather than pressing one of the four buttons) by looping to display the menu until the user presses one of the buttons.
- Return the integer value corresponding to the button press.

The algorithm for the *explaine* function is:

- Print an explanation of *e*, the **exp** function, and how to find approximate values.

The algorithm for the *limite* function is:

- Call a subfunction *askforn* to prompt the user for an integer *n*.
- Calculate and print the approximate value of *e* using *n*.

The algorithm for the subfunction *askforn* is:

- Prompt the user for a positive integer for *n*.
- Loop to print an error message and reprompt until the user enters a positive integer.
- Return the positive integer *n*.

The algorithm for the *expfn* function is:

- Receives the value of *x* as an input argument.
- Prints the value of **exp(x)**.
- Assigns an arbitrary value for the number of terms *n* (an alternative method would be to prompt the user for this).
- Call a subfunction *appex* to find an approximate value of **exp(x)** using a series with *n* terms.
- Prints this approximate value.

The algorithm for the *appex* subfunction is:

- Receives *x* and *n* as input arguments.
- Initializes a variable for the running sum of the terms in the series (to 1 for the first term) and for a running product, which will be the factorials in the denominators.
- Loops to add the *n* terms to the running sum.
- Returns the resulting sum.

The entire program consists of the following M-file script and four M-file functions:

eapplication.m

```
% This script explores e and the exponential function

% Call a function to display a menu and get a choice
choice = eoption;
```

(Continued)

```matlab
% Choice 4 is to exit the program
while choice ~= 4
    switch choice
        case 1
            % Explain e
            explaine;
        case 2
            % Approximate e using a limit
            limite;
        case 3
            % Approximate exp(x) and compare to exp
            x = input('Please enter a value for x: ');
            expfn(x);
    end
    % Display menu again and get user's choice
    choice = eoption;
end
```

eoption.m

```matlab
function choice = eoption
% Print the menu of options and error-check
% until the user pushes one of the buttons
choice = menu('Choose an e option', 'Explanation', ...
    'Limit', 'Exponential function', 'Exit Program');
% If the user closes the menu box rather than
% pushing one of the buttons, choice will be 0
while choice == 0
    disp('Error - please choose one of the options.')
    choice = menu('Choose an e option', 'Explanation', ...
        'Limit', 'Exponential function', 'Exit Program');
end
```

explaine.m

```
function explaine
% This function explains a little bit about e
fprintf('The constant e is called the natural')
fprintf(' exponential base.\n')
fprintf('It is used extensively in mathematics and')
fprintf(' engineering.\n')
fprintf('The value of the constant e is ~ 2.1718\n')
fprintf('Raising e to the power of x is so common that')
fprintf(' this is called the exponential function.\n')
fprintf('An approximation for e is found using a limit.\n')
fprintf('An approximation for the exponential function')
fprintf(' can be found using a series.\n')
```

limite.m

```
function limite
% Approximates e using a limit
% Call a subfunction to prompt user for n
n = askforn;
fprintf('An approximation of e with n = %d is %.2f\n', ...
    n, (1 + 1/n) ^ n)

function outn = askforn
% This subfunction prompts the user for n
% It error-checks to make sure n is a positive integer
inputnum = input('Enter a positive integer for n: ');
num2 = int32(inputnum);
while num2 ~= inputnum || num2 < 0
    inputnum = input('Invalid! Enter a positive integer: ');
    num2 = int32(inputnum);
end
outn = inputnum;
```

expfn.m

```
function expfn(x)
% Compares the built-in function exp and a
% series approximation
fprintf('Value of built-in exp(x) is %.2f\n',exp(x))
% n is arbitrary number of terms
n = 10;
fprintf('Approximate exp(x) is %.2f\n', appex(x,n))
function outval = appex(x,n)
% Approximates e to the x power using terms up to
% x to the nth power
% Initialize the running sum to 1 (for the first term)
outval = 1;
runprod = 1;
for i = 1:n
    runprod = runprod * i;
    outval = outval + (x^i)/runprod;
end
```

Running the script will bring up the menu seen in Figure 5.1.

Then, what happens will depend on which button(s) the user pushes. Every time the user pushes a button, the appropriate function will be called and then this menu will appear again. This will continue until the user presses the button **Exit Program**. Examples will be given here of running the script, with different sequences of button presses.

FIGURE 5.1

Menu Figure Window for eapplication program.

In the following example, the user

- Closed the menu window, which caused the error message and brought up a new menu.

- Chose **Explanation**.

- Chose **Exit Program**.

```
>> eapplication
Error - please choose one of the options.
The constant e is called the natural exponential base.
It is used extensively in mathematics and engineering.
The value of the constant e is ~ 2.1718
```

```
Raising e to the power of x is so common that this is
called the exponential function.
An approximation for e is found using a limit.
An approximation for the exponential function can be found
using a series.
```

In the following example, the user

- Chose **Limit**;

  - when prompted for *n*, entered two invalid values before finally entering a valid positive integer.

- Chose **Exit Program**.

```
>> eapplication
Enter a positive integer for n: -4
Invalid! Enter a positive integer: 5.5
Invalid! Enter a positive integer: 10
An approximation of e with n = 10 is 2.59
```

In order to see the difference in the approximate value for *e* as n increases, the user kept choosing **Limit** and entering larger and larger values each time in the following example:

```
>> eapplication
Enter a positive integer for n: 4
An approximation of e with n = 4 is 2.44
Enter a positive integer for n: 10
An approximation of e with n = 10 is 2.59
Enter a positive integer for n: 30
An approximation of e with n = 30 is 2.67
Enter a positive integer for n: 100
An appoximation of e with n = 100 is 2.70
```

In the following example, the user

- Chose **Exponential function**;

  - when prompted, entered 4.6 for *x*.

- Chose **Exponential function** again;

  - when prompted, entered –2.3 for *x*.

- Chose **Exit Program**.

```
>> eapplication
Please enter a value for x: 4.6
Value of built-in exp(x) is 99.48
Approximate exp(x) is 98.71
```

```
Please enter a value for x: -2.3
Value of built-in exp(x) is 0.10
Approximate exp(x) is 0.10
```

5.4 VARIABLE SCOPE

The *scope* of any variable is the workspace in which it is valid. The workspace created in the Command Window is called the *base workspace*.

As we have seen before, if a variable is defined in any function it is a *local variable* to that function, which means that it is known and used only within that function. For example, in the following function that calculates the sum of the elements in a vector, there is a local loop variable *i*.

mysum.m

```
function runsum = mysum(vec)
% This function sums a vector
runsum = 0;
for i=1:length(vec)
   runsum = runsum + vec(i);
end
```

Running this function does not add any variables to the workspace, as demonstrated:

```
>> clear
>> who
>> disp(mysum([5 9 1]))
   15
>> who
>>
```

Also, variables that are defined in the Command Window cannot be used in a function.

However, scripts (as opposed to functions) *do* interact with the variables that are defined in the Command Window. For example, the function is changed to be a script *mysummfile*.

mysummfile.m

```
% This script sums a vector
vec = 1:5;
runsum = 0;
```

(Continued)

```
for i=1:length(vec)
    runsum = runsum + vec(i);
end
disp(runsum)
```

The variables defined in the script do become part of the workspace:

```
>> clear
>> who
>> mysummfile
    15
>> who
Your variables are:
i runsum vec
```

Variables that are defined in the Command Window can be used in a script, but cannot be used in a function. For example, the vector *vec* could be defined in the Command Window (instead of in the script), but then used in the script:

mysummfileii.m

```
% This script sums a vector from the Command Window
runsum = 0;
for i=1:length(vec)
    runsum = runsum + vec(i);
end
```

```
>> clear
>> vec = 1:7;
>> who
Your variables are:
vec
>> mysummfileii
>> who
Your variables are:
i runsum vec
>> runsum
runsum =
    28
```

This, however, is very poor programming style. It is much better to pass the vector *vec* to a function.

It is possible, in MATLAB as well in other languages, to have *global variables* that can be shared by functions. Although there are some cases in which using global variables is efficient, it is generally regarded as poor programming style.

5.4.1 Persistent Variables

Normally, when a function stops executing, the local variables from that function are cleared. That means that every time a function is called, memory is allocated and used while the function is executing, but released when it ends. With variables that are declared as **persistent variables**, however, the value is not cleared so the next time the function is called, the variable still exists and retains its former value.

The following program demonstrates this. The script calls a function *func1*, which initializes a variable *count* to 0, then increments it, and then prints the value. Every time this function is called, the variable is created, initialized to 0, changed to 1, and then cleared when the function exits. The script then calls a function *func2*, which first declares a persistent variable *count*. If the variable has not yet been initialized, which will be the case the first time the function is called, it is initialized to 0. Then, like the first function, the variable is incremented and the value is printed. With the second function, however, when the function exits, the variable remains with its value, so the next time the function is called the variable is incremented again.

persistex.m

```
% This script demonstrates persistent variables
% The first function has a variable count
fprintf('This is what happens with a normal variable:\n')
func1
func1
% The second fn has a persistent variable count
fprintf('\nThis is what happens with a persistent variable:\n')
func2
func2
```

func1.m

```
function func1
% This function increments a variable count
count = 0;
count = count + 1;
fprintf('The value of count is %d\n',count)
```

func2.m

```
function func2
% This function increments a persistent variable count
persistent count
if isempty(count)
    count = 0;
end
count = count + 1;
fprintf('The value of count is %d\n',count)
```

The functions can be called from the script or from the Command Window, as shown here. For example, the functions are called first from the script. With the persistent variable, the value of *count* is incremented. Then, *func1* is called from the Command Window, and *func2* is also called from the Command Window. Since the value of the persistent variable had the value 2, this time it is incremented to 3.

```
>> persistex
This is what happens with a normal variable:
The value of count is 1
The value of count is 1
This is what happens with a persistent variable:
The value of count is 1
The value of count is 2
>> func1
The value of count is 1
>> func2
The value of count is 3
```

As can be seen here, every time the function *func1* is called, whether from *persistex* or from the Command Window, the value of 1 is printed. However, with *func2* the variable count is incremented every time it is called. It is first called in this example from *persistex* twice, so count is 1 and then 2. Then, when called from the Command Window, it is incremented to 3.

Of course, the numbering of the error messages will continue if the function is executed again.

5.5 DEBUGGING TECHNIQUES

Any error in a computer program is called a **bug**. This term is thought to date back to the 1940s, when a problem with an early computer was found

PRACTICE 5.6

The following function *posnum* prompts the user to enter a positive number and loops to error-check. It returns the positive number entered by the user. It calls a subfunction in the loop to print an error message. The subfunction has a persistent variable to count the number of times an error has occurred. Here is an example of calling the function:

```
>> enteredvalue = posnum
Enter a positive number: -5
Error # 1 ... Follow instructions!
Does -5.00 look like a positive number to you?
Enter a positive number: -33
Error # 2 ... Follow instructions!
Does -33.00 look like a positive number to you?
Enter a positive number: 6
enteredvalue =
     6
```

Fill in the following subfunction to accomplish this.

posnum.m

```
function num = posnum
% Prompt user and error-check until the
% user enters a positive number
num = input('Enter a positive number: ');
while num < 0
    errorsubfn(num)
    num = input('Enter a positive number: ');
end

function errorsubfn(num)
```

to have been caused by a moth in the computer's circuitry! The process of finding errors in a program, and correcting them, is still called *debugging*.

5.5.1 Types of Errors

There are several different kinds of errors that can occur in a program, which fall into the categories of *syntax errors*, *run-time errors*, and *logical errors*.

Syntax errors are mistakes in using the language. Examples of syntax errors are missing a comma or a quotation mark, or misspelling a word. MATLAB itself will flag syntax errors and give an error message. For example, the following string is missing the end quote:

```
>> mystr = 'how are you;
??? mystr = 'how are you;
                    |
Error: A MATLAB string constant is not terminated
properly.
```

Another common mistake is to spell a variable name incorrectly, which MATLAB will also catch.

```
>> value = 5;
>> newvalue = valu + 3;
??? Undefined function or variable 'valu'.
```

Run-time, or execution-time, errors are found when a script or function is executing. With most languages, an example of a run-time error would be attempting to divide by zero. However, in MATLAB, this will generate a warning message. Another example would be attempting to refer to an element in an array that does not exist.

runtime_ex.m

```
% This script shows an execution-time error
vec = 3:5;
for i = 1:4
    disp(vec(i))
end
```

This script initializes a vector with three elements, but then attempts to refer to a fourth. Running it prints the three elements in the vector, and then an error message is generated when it attempts to refer to the fourth element. Notice that it gives an explanation of the error, and it gives the line number in the script in which the error occurred.

```
>> runtime_ex
     3
     4
     5
??? Attempted to access vec(4); index out of bounds because
numel(vec)=3.
Error in ==> runtime_ex at 6
        disp(vec(i))
```

Logical errors are more difficult to locate, because they do not result in any error message. A logical error is a mistake in reasoning by the programmer, but it is not a mistake in the programming language. An example of a logical error would be dividing by 2.54 instead of multiplying in order to convert inches to centimeters. The results printed or returned would be incorrect, but this might not be obvious.

All programs should be robust and should wherever possible anticipate potential errors, and guard against them. For example, whenever there is input into a program, the program should error-check and make sure that the input is in the correct range of values. Also, before dividing, the denominator should be checked to make sure that it is not zero.

Despite the best precautions, there are bound to be errors in programs.

5.5.2 Tracing

Many times, when a program has loops and/or selection statements and is not running properly, it is useful in the debugging process to know exactly which statements have been executed. For example, here is a function that attempts to display In Middle Of Range if the argument passed to it is in the range from 3 to 6, and Out Of Range otherwise.

testifelse.m

```
function testifelse(x)
% This function will test the debugger
if 3 < x < 6
    disp('In middle of range')
else
    disp('Out of range')
end
```

However, it seems to print In Middle Of Range for all values of x:

```
>> testifelse(4)
In middle of range
>> testifelse(7)
In middle of range
>> testifelse(-2)
In middle of range
```

One way of following the flow of the function, or *tracing* it, is to use the **echo** function. The **echo** function, which is a toggle, will display every statement as it is executed as well as results from the code. For scripts, just **echo** can be typed, but for functions, the name of the function must be specified, for example,

```
echo functionname on/off
>> echo testifelse on
```

```
>> testifelse(-2)
% This function will test the debugger
if 3 < x < 6
    disp('In middle of range')
In middle of range
end
```

We can see from this result that the action of the **if** clause was executed.

5.5.3 Editor/Debugger

MATLAB has many useful functions for debugging, and debugging can also be done through its editor, called the Editor/Debugger.

Typing **help debug** at the prompt in the Command Window will show some of the debugging functions. Also, in the Help Browser, clicking the Search tab and then typing **debugging** will display basic information about the debugging processes.

It can be seen in the previous example that the action of the **if** clause was executed and it printed In Middle Of Range, but just from that it cannot be determined why this happened. There are several ways to set *breakpoints* in a file (script or function) so that the variables or expressions can be examined. These can be done from the Editor/Debugger, or commands can be typed from the Command Window. For example, the following **dbstop** command will set a breakpoint in the fifth line of this function (which is the action of the **if** clause), which allows us to type variable names and/or expressions to examine their values at that point in the execution. The function **dbcont** can be used to continue the execution, and **dbquit** can be used to quit the debug mode. Notice that the prompt becomes K>> in debug mode.

```
>> dbstop testifelse 5
>> testifelse(-2)
5 disp('In middle of range')
K>> x
x =
    -2
K>> 3 < x
ans =
    0
K>> 3 < x < 6
ans =
    1
K>> dbcont
In middle of range
end
>>
```

By typing the expressions 3 < x and then 3 < x < 6, we can determine that the expression 3 < x will return either 0 or 1. Both 0 and 1 are less than 6, so the expression will always be true, regardless of the value of x!

5.5.4 Function Stubs

Another common debugging technique, which is used when there is a script main program that calls many functions, is to use *function stubs*. A function stub is a placeholder, used so that the script will work even though that particular function hasn't been written yet. For example, a programmer might start with a script main program that consists of calls to three function that accomplish all the tasks.

mainmfile.m

```
% This program gets values for x and y, and
%     calculates and prints z
[x, y] = getvals;
z = calcz(x,y);
printall(x,y,z)
```

The three functions have not yet been written, however, so function stubs are put in place so that the script can be executed and tested. The function stubs consist of the proper function headers, followed by a simulation of what the function will eventually do (e.g., it puts arbitrary values in for the output arguments).

getvals.m

```
function [x, y] = getvals
x = 33;
y = 11;
```

calcz.m

```
function z = calcz(x,y)
z = 2.2;
```

printall.m

```
function printall(x,y,z)
disp('Something')
```

Then, the functions can be written and debugged one at a time. It is much easier to write a working program using this method than to attempt to write everything at once—then, when errors occur, it is not always easy to determine where the problem is!

SUMMARY

Common Pitfalls

- Not matching up arguments in a function call with the input arguments in a function header
- Not having enough variables in an assignment statement to store all the values returned by a function through the output arguments
- Attempting to call a function that does not return a value from an assignment statement, or from an output statement
- Not using the same name for the function and the file in which it is stored
- Not thoroughly testing functions for all possible inputs and outputs
- Forgetting that persistent variables are updated every time the function in which they are declared is called, whether from a script or from the Command Window

Programming Style Guidelines

- If a function is calculating one or more values, return these value(s) from the function by assigning them to output variable(s).
- Give the function and the file in which it is stored the same name.
- Function headers and function calls must correspond. The number of arguments passed to a function must be the same as the number of input arguments in the function header. If the function returns values, the number of variables in the left-hand side of an assignment statement should match the number of output arguments returned by the function.
- If arguments are passed to the function in the function call, do not replace these values by using **input** in the function itself.
- Functions that calculate and return value(s) will not normally also print them.
- Functions should not normally be longer than one page in length.
- Do not declare variables in the Command Window and then use them in a script, or vice versa.
- Pass all values to be used in functions to input arguments in the functions.
- When writing large programs with many functions, start with the main program script and use function stubs, filling in one function at a time while debugging.

MATLAB Functions and Commands	
echo	dbcont
dbstop	dbquit

Exercises

1. Write a function that will receive as an input argument a temperature in degrees Fahrenheit, and will return the temperature in both degrees Celsius and Kelvin. The conversion factors are: C = (F − 32) * 5/9 and K = C + 273.15.

2. Write a function that will receive as an input argument a length in feet and will return the length in both yards and centimeters. One yard is equal to 3 feet. One inch is equal to 2.54 centimeters, and there are 12 inches in a foot.

3. A vector can be represented by its rectangular coordinates x and y or by its polar coordinates r and θ. The conversions from polar to rectangular are x = r cos(θ) and y = r sin(θ). Write a function to receive the polar coordinates as input arguments and return the corresponding rectangular coordinates.

4. A vector can be represented by its rectangular coordinates x and y or by its polar coordinates r and θ. For positive values of x and y, the conversions from rectangular to polar coordinates in the range from 0 to 2 π are $r = \sqrt{x^2 + y^2}$ and θ = arctan(y/x). Write a function to receive as input arguments the rectangular coordinates and return the corresponding polar coordinates.

5. Write a function to calculate the volume and surface area of a hollow cylinder. It receives as input arguments the radius of the cylinder base and the height of the cylinder. The volume is given by π r^2 h, and the surface area is 2 π r h.

6. Hurricanes are categorized based on the winds. The following table shows the category number for hurricanes with varying wind ranges and what the storm surge is (in feet above normal).

1	74–95	4–5
2	96–110	6–8
3	111–130	9–12
4	131–155	13–18
5	>155	>18

Write a function that will receive as an input argument the wind speed, and will return the category number and the minimum value of the typical storm surge.

7. Write a function that will receive the radius of a circle and will print both the radius and diameter of the circle in a sentence format. This function will not return any value; it simply prints.

8. Write a function that will receive as input arguments a length in inches, and will print in sentence format the length in both inches and centimeters (1 inch = 2.54 centimeters). Note that this function will not return any value.

9. Write a function that will receive an integer n and a character as input arguments, and will print the character n times.

10. Convert the *printstars* script from Chapter 4 to a function that receives as inputs the number of rows and columns, and prints a box of asterisks with the specified number of rows and columns.

11. Convert the *multtable* function from Chapter 4 to a function that receives as input arguments the number of rows and columns and prints a multiplication table (rather than returning the matrix).

12. Write a function that will receive a matrix as an input argument, and prints it in a table format.

13. Write a function that receives a matrix as an input argument, and prints a random row from the matrix.

14. Write a function that receives a count as an input argument, and prints the value of the count in a sentence that would read "It happened 1 time." if the value of the count is 1, or "It happened xx times." if the value of count (xx) is greater than 1.

15. Write a function that will print an explanation of temperature conversions. The function does not receive any input arguments; it simply prints.

16. Write a function that prompts the user for a value of an integer n, and returns the value of n. No input arguments are passed to this function.

17. Write a function that prompts the user for a value of an integer n, and returns a vector of values from 1 to n. The function should error-check to make sure that the user enters an integer. No input arguments are passed to this function.

18. Write a script that will

   ▪ Call a function to prompt the user for an angle in degrees
   ▪ Call a function to calculate and return the angle in radians
     (note: π radians = 180°
   ▪ Call a function to print the result

   Note that the solution to this problem involves four M-files: one that acts as a main program, and three for the functions.

19. Modify the preceding program so that the function to calculate the angle is a subfunction to the function that prints.

20. Write a program to calculate and print the area and circumference of a circle. There should be one script and three functions to accomplish this (one that prompts for the radius, one that calculates the area and circumference, and one that prints).

21. The lump sum S to be paid when interest on a loan is compounded annually is given by $S = P(1 + i)^n$, where P is the principal invested, i is the interest rate, and n is the number of years. Write a program that will plot the amount S as it increases through the years from 1 to n. The main script will call a function to prompt the user for the number of years (and error-check to make sure that the user enters a positive integer). The script will then call a function that will plot S for years 1 through n. It will use 0.05 for the interest rate and $10,000 for P.

22. Write a program to write a length conversion chart to a file. It will print lengths in feet, from 1 to an integer specified by the user in one column, and the corresponding length in meters (1 foot = 0.3048 meters) in a second column. The main script will call one function that prompts the user for the maximum length in feet; this function must error-check to make sure that the user enters a valid positive integer. The script then calls a function to write the lengths to a file.

23. The resistance R in ohms of a conductor is given by $R = \dfrac{E}{I}$, where E is the potential in volts and I is the current in amperes. Write a script that will

 ■ Call a function to prompt the user for the potential and the current.
 ■ Call a function that will print the resistance; this will call a subfunction to calculate and return the resistance.

24. The power in watts is given by P = EI. Modify the preceding program to calculate and print both the resistance and the power. Modify the subfunction so that it calculates and returns both values.

25. The distance between any two points (x_1, y_1) and (x_2, y_2) is given by:

$$distance = \sqrt{(x_1 - x_2)^2 + (y_1 - y_2)^2}$$

The area of a triangle is:

$$area = \sqrt{s * (s - a) * (s - b) * (s - c)}$$

where *a, b,* and *c* are the lengths of the sides of the triangle, and *s* is equal to half the sum of the lengths of the three sides of the triangle. Write a script that will prompt the user to enter the coordinates of three points that determine a triangle (e.g., the x and y coordinates of each point). The script will then calculate and print the area of the triangle. It will call one function to calculate the area of the triangle. This function will call a subfunction, which calculates the length of the side formed by any two points (the distance between them).

26. Write a program to write a temperature conversion chart to a file. The main script will:

 ■ Call a function that explains what the program will do.
 ■ Call a function to prompt the user for the minimum and maximum temperatures in degrees Fahrenheit, and return both values. This function checks to make sure that the minimum is less than the maximum, and calls a subfunction to swap the values if not.
 ■ Call a function to write temperatures to a file: the temperature in degrees F from the minimum to the maximum in one column, and the corresponding temperature in degrees Celsius in another column. The conversion is C = (F − 32) * 5/9.

27. Write a script that will ask the user to choose his or her favorite science class, and print a message regarding that course. It will call a function to display a menu of choices (using the menu function); this function will error-check to

make sure that the user presses one of the buttons. The function will return the number corresponding to the user's choice. The script will then print a message.

28. Write a menu-driven program to convert a time in seconds to other units (minutes, hours, etc.). The main script will loop to continue until the user chooses to exit. Each time in the loop, the script will generate a random time in seconds, call a function to present a menu of options, and print the converted time. The conversions must be made by individual functions (e.g., one to convert from seconds to minutes). All user entries must be error-checked.

29. Write a menu-driven program to investigate the constant π. Model it after the program that explores the constant e. Pi (π) is the ratio of a circle's circumference to its diameter. Many mathematicians have found ways to approximate π. For example, Machin's formula is:

$$\frac{\pi}{4} = 4 \arctan\left(\frac{1}{5}\right) - \arctan\left(\frac{1}{239}\right)$$

Leibniz found that π can be approximated by:

$$\pi = \frac{4}{1} - \frac{4}{3} + \frac{4}{5} - \frac{4}{7} + \frac{4}{9} - \frac{4}{11} + \cdots$$

This is called a sum of a series. There are six terms shown in this series. The first term is 4, the second term is –4/3, the third term is 4/5, and so forth. For example, the menu-driven program might have the following options:

- Print the result from Machin's formula.
- Print the approximation using Leibniz' formula, allowing the user to specify how many terms to use.
- Print the approximation using Leibniz' formula, looping until a good approximation is found.
- Exit the program.

30. Modify the function *func2* from the text that has a persistent variable count. Instead of having the function print the value of count, the value should be returned.

31. Write a function *per2* that receives one number as an input argument. The function has a persistent variable that sums the values passed to it. Here are the first two times the function is called:

```
>> per2(4)
ans =
    4
>> per2(6)
ans =
    10
```

32. What would be the output from the following program? Think about it, write down your answer, and then type it in to verify.

testscope.m

```
answer = 5;
fprintf('Answer is %d\n',answer)
pracfn
pracfn
fprintf('Answer is %d\n',answer)
printstuff
fprintf('Answer is %d\n',answer)
```

pracfn.m

```
function pracfn
persistent count
if isempty(count)
    count = 0;
end
count = count + 1;
fprintf('This function has been called %d times.\n',count)
```

printstuff.m

```
function printstuff
answer = 33;
fprintf('Answer is %d\n',answer)
pracfn
fprintf('Answer is %d\n',answer)
```

String Manipulation

CONTENTS

A *string* in the MATLAB® software consists of any number of characters and is contained in single quotes. Actually, strings are treated as vectors in which every element is a single character, which means that many of the vector operations and functions that we have seen already work with strings. MATLAB also has many built-in functions that are written specifically to manipulate strings. In some cases, strings contain numbers, and it is useful to convert from strings to numbers and vice versa; MATLAB has functions to do this, also.

There are many applications for using strings, even in fields that are predominantly numerical. For example, when data files consist of combinations of numbers and characters, it is often necessary to read each line from the file as a string, break the string into pieces, and convert the parts that contain numbers to number variables that can be used in computations. In this chapter the string manipulation techniques necessary for this will be introduced, and in Chapter 8 applications in file input/output will be demonstrated.

6.1 CREATING STRING VARIABLES

A string consists of any number of characters (including, possibly, none). These are examples of strings:

```
' '
'x'
'cat'
'Hello there'
'123'
```

A *substring* is a subset or part of a string. For example, 'there' is a substring within the string 'Hello there'.

Characters include letters of the alphabet, digits, punctuation marks, white space, and control characters. *Control characters* are characters that cannot be printed, but accomplish a task (such as a backspace or tab). *Whitespace characters* include the space, tab, newline (which moves the cursor down to the next line), and carriage return (which moves the cursor to the beginning of the current line). *Leading blanks* are blank spaces at the beginning of a string, for example, '   hello', and *trailing blanks* are blank spaces at the end of a string.

There are several ways that string variables can be created. One is using assignment statements:

```
>> word = 'cat';
```

Another method is to read into a string variable. Recall that to read into a string variable using the **input** function, the second argument 's' must be included:

```
>> strvar = input('Enter a string: ', 's')
Enter a string: xyzabc
strvar =
xyzabc
```

If leading or trailing blanks are typed by the user, these will be stored in the string. For example, in the following the user entered four blanks and then 'xyz':

```
>> s = input('Enter a string: ','s')
Enter a string:     xyz
s =
    xyz
```

6.1.1 Strings as Vectors

Strings are treated as *vectors of characters*—or in other words, a vector in which every element is a single character—so many vector operations can be performed. For example, the number of characters in a string can be found using the **length** function:

```
>> length('cat')
ans =
      3
>> length(' ')
ans =
      1
>> length('')
ans =
      0
```

Notice that there is a difference between an **empty string**, which has a length of zero, and a string consisting of a blank space, which has a length of one.

Expressions can refer to an individual element (a character within the string), or a subset of a string or a transpose of a string:

```
>> mystr = 'Hi';
>> mystr(1)
ans =
H
>> mystr'
ans =
H
i
>> sent = 'Hello there';
>> length(sent)
ans =
     11
>> sent(4:8)
ans =
lo th
```

Notice that the blank space in the string is a valid character within the string.

A matrix can be created, which consists of strings in each row. So, essentially it is created as a column vector of strings, but the end result is that this would be treated as a matrix in which every element is a character:

```
>> wordmat = ['Hello';'Howdy']
wordmat =
Hello
Howdy
>> size(wordmat)
ans =
     2     5
```

This created a 2 × 5 matrix of characters.

With a character matrix, we can refer to an individual element, which is a character, or an individual row, which is one of the strings:

```
>> wordmat(2,4)
ans =
d
>> wordmat(1,:)
ans =
Hello
```

Since rows within a matrix must always be the same length, the shorter strings must be padded with blanks so that all strings have the same length, otherwise an error will occur.

```
>> greetmat = ['Hello'; 'Goodbye']
??? Error using ==> vertcat
CAT arguments dimensions are not consistent.
>> greetmat = ['Hello  '; 'Goodbye']
greetmat =
Hello
Goodbye
>> size(greetmat)
ans =
2     7
```

 PRACTICE 6.1

Prompt the user for a string. Print the length of the string and also the last character in the string. Make sure that this works regardless of what the user enters.

6.2 OPERATIONS ON STRINGS

MATLAB has many built-in functions that work with strings. Some of the string manipulation functions that perform the most common operations will be described here.

6.2.1 Concatenation

String concatenation means to join strings together. Of course, since strings are just vectors of characters, the method of concatenating vectors works for strings, also. For example, to create one long string from two strings, it is possible to join them by putting them in square brackets:

```
>> first = 'Bird';
>> last = 'house';
>> [first last]
ans =
Birdhouse
```

The function **strcat** does this also horizontally, meaning that it creates one longer string from the inputs.

```
>> first = 'Bird';
>> last = 'house';
>> strcat(first,last)
ans =
Birdhouse
```

There is a difference between these two methods of concatenating, however, if there are leading or trailing blanks in the strings. The method of using the square brackets will concatenate the strings, including all leading and trailing blanks.

```
>> str1 = 'xxx    ';
>> str2 = '    yyy';
>> [str1 str2]
ans =
xxx        yyy
>> length(ans)
ans =
      12
```

The **strcat** function, however, will remove trailing blanks (but not leading blanks) from strings before concatenating. Notice that in these examples, the trailing blanks from *str1* are removed, but the leading blanks from *str2* are not:

```
>> strcat(str1,str2)
ans =
xxx   yyy
>> length(ans)
ans =
     9
>> strcat(str2,str1)
ans =
   yyyxxx
>> length(ans)
ans =
     9
```

The function **strvcat** will concatenate vertically, meaning that it will create a column vector of strings.

```
>> strvcat(first,last)
ans =
Bird
house
>> size(ans)
ans =
     2      5
```

Note that **strvcat** will pad with extra blanks automatically, in this case to make both strings have a length of 5.

 PRACTICE 6.2

Create the following string variables:

```
v1 = 'Mechanical';
v2 = 'Engineering';
```

Then, get the length of each string. Create a new variable, *v3*, which is a substring of *v2* that stores just 'Engineer'. Create a matrix consisting of the values of *v1* and *v2* in separate rows.

6.2.2 Creating Customized Strings

There are several built-in functions that create customized strings, including **char, blanks**, and **sprintf**.

We have seen already that the **char** function can be used to convert from an ASCII code to a character, for example:

```
>> char(97)
ans =
a
```

The **char** function can also be used to create a matrix of characters. When using the **char** function to create a matrix, it will automatically pad the strings within the rows with blanks as necessary so that they are all the same length, just like **strvcat**.

```
>> clear greetmat
>> greetmat = char('Hello','Goodbye')
greetmat =
Hello
Goodbye
>> size(greetmat)
ans =
     2     7
```

The **blanks** function will create a string consisting of n blank characters—which are kind of hard to see here! However, in MATLAB if the mouse is moved to highlight the result in *ans*, the blanks can be seen.

```
>> blanks(4)
ans =

>> length(ans)
ans =
     4
```

Usually this function is most useful when concatenating strings, and you want a number of blank spaces in between. For example, this will insert five blank spaces in between the words:

```
>> [first blanks(5) last]
ans =
Bird     house
```

Displaying the transpose of the **blanks** function can also be used to move the cursor down. In the Command Window, it would look like this:

```
>> disp(blanks(4)')


>>
```

This is useful in a script or function to create space in output, and is essentially equivalent to printing the newline character four times.

The **sprintf** function works exactly like the **fprintf** function, but instead of printing it creates a string. Here are several examples in which the output is not suppressed so the value of the string variable is shown:

```
>> sent1 = sprintf('The value of pi is %.2f', pi)
sent1 =
The value of pi is 3.14
>> sent2 = sprintf('Some numbers: %5d, %2d', 33, 6)
sent2 =
Some numbers:    33,  6
>> length(sent2)
ans =
    23
```

In the following example, on the other hand, the output of the assignment is suppressed so the string is created including a random integer and stored in the string variable. Then, some exclamation points are concatenated to that string.

```
>> phrase = sprintf('A random integer is %d', ...
        randint(1,1,[5,10]));
>> strcat(phrase, '!!!')
ans =
A random integer is 7!!!
```

All the conversion specifiers that can be used in the **fprintf** function can also be used in the **sprintf** function.

6.2.2.1 *Applications of Customized Strings: Prompts, Labels, Arguments to Functions*

One very useful application of this is to include numbers in strings, which are used for plot titles and axis labels. For example, assume that a file 'expnoanddata.dat' stores an experiment number, followed by the experiment data. In this case the experiment number is 123, and then the rest of the file consists of the actual data.

```
123  4.4    5.6    2.5    7.2    4.6    5.3
```

The following script would load this data and plot it with a title that includes the experiment number.

```
plotexpno.m
% This script loads a file that stores an experiment #
% followed by the actual data. It plots the data and puts
% the experiment # in the plot title
load expnoanddata.dat
exper_no = expnoanddata(1);
data = expnoanddata(2:end);
plot(data,'ko')
title(sprintf('Data from experiment %d', exper_no))
```

FIGURE 6.1

*Customized title in plot
using* **sprintf.**

The script loads all numbers from the file into a row vector. It then separates the vector; it stores the first element, which is the experiment number in a variable *exper_no*, and the rest of the vector in a variable *data* (the rest being from the second element to the end). It then plots the data, using **sprintf** to create the title that includes the experiment number as seen in Figure 6.1.

 PRACTICE 6.3

In a loop, create and print strings with filenames *file1.dat*, *file2.dat*, and so on for file numbers 1 through 5.

QUICK QUESTION!

How could we use the **sprintf** function in customizing prompts for the **input** function?

Answer: For example, if you want the contents of a string variable printed in a prompt, **sprintf** can be used:

```
>> username = input('Please enter your name: ', 's');
Please enter your name: Bart
>> prompt = sprintf('%s, Enter your id #: ',username);
>> id_no = input(prompt)
Bart, Enter your id #: 177
id_no =
    177
```

Another way of accomplishing this (in a script or function) would be:

```
fprintf('%s, Enter your id #: ',username);
id_no = input('');
```

Notice that the calls to the **sprintf** and **fprintf** functions are identical except that the **fprintf** prints (so there is no need for a prompt in the **input** function) whereas the **sprintf** creates a string that can then be displayed by the **input** function. In this case using **sprintf** seems cleaner than using **fprintf** and then having an empty string for the prompt in **input**.

As another example, the following program prompts the user for endpoints (x_1, y_1) and (x_2, y_2) of a line segment, and calculates the midpoint of the line segment, which is the point (x_m, y_m). The coordinates of the midpoint are found by:

$$x_m = \frac{1}{2}(x_1 + x_2) \qquad y_m = \frac{1}{2}(y_1 + y_2)$$

The script *midpoint.m* calls a function *entercoords* to separately prompt the user for the x- and y-coordinates of the two endpoints, calls a function *findmid* that calculates separately the x- and y-coordinates of the midpoint, and then prints this midpoint. When the program is executed, the output looks like this:

```
>> midpoint
Enter the x coord of the first endpoint: 2
Enter the y coord of the first endpoint: 4
Enter the x coord of the second endpoint: 3
Enter the y coord of the second endpoint: 8
The midpoint is (2.5, 6.0)
```

In this example, the word 'first' or 'second' is passed to the **entercoords** function so that it can use whichever word is passed in the prompt. The **sprintf** function generates this customized prompt.

midpoint.m

```
% This program finds the midpoint of a line segment
[x1, y1] = entercoords('first');
[x2, y2] = entercoords('second');
midx = findmid(x1,x2);
midy = findmid(y1,y2);
fprintf('The midpoint is (%.1f, %.1f)\n',midx,midy)
```

entercoords.m

```
function [xpt, ypt] = entercoords(word)
% Prompts the user for the coordinates of an endpoint

% Two different methods are used to customize the
% prompt to show the difference
fprintf('Enter the x coord of the %s endpoint: ', word)
xpt = input('');
prompt = sprintf('Enter the y coord of the %s endpoint: ', ...
    word);
ypt = input(prompt);
```

findmid.m

```
function mid = findmid(pt1,pt2)
% Calculate a coordinate (x or y) of the
% midpoint of a line segment
mid = 0.5 * (pt1 + pt2);
```

6.2.3 Removing Whitespace Characters

MATLAB has functions that will remove trailing blanks from the end of a string and/or leading blanks from the beginning of a string.

The **deblank** function will remove blank spaces from the end of a string. For example, if some strings are padded in a string matrix so that all are the same length, it is frequently preferred to then remove those extra blank spaces in order to actually use the string.

```
>> names = char('Sue', 'Cathy','Xavier')
names =
Sue
Cathy
Xavier
>> name1 = names(1,:)
name1 =
Sue
>> length(name1)
ans =
    6
>> name1 = deblank(name1);
>> length(name1)
ans =
    3
```

Note: The **deblank** function removes only trailing blanks from a string, not leading blanks.

The **strtrim** function will remove both leading and trailing blanks from a string, but not blanks in the middle of the string. In the following example, the three blanks in the beginning and four blanks in the end are removed, but not the two blanks in the middle. Selecting the result in MATLAB with the mouse would show the blank spaces.

```
>> strvar = [blanks(3) 'xx' blanks(2) 'yy' blanks(4)]
strvar =
   xx  yy
>> length(strvar)
ans =
    13
>> strtrim(strvar)
ans =
xx  yy
>> length(ans)
ans =
    6
```

6.2.4 Changing Case

MATLAB has two functions that convert strings to all uppercase letters, or all lowercase, called **upper** and **lower**.

```
>> mystring = 'AbCDEfgh';
>> lower(mystring)
ans =
abcdefgh
>> upper(ans)
ans =
ABCDEFGH
```

6.2.5 Comparing Strings

There are several functions that compare strings and return logical true if they are equivalent, or logical false if not. The function **strcmp** compares strings, character by character. It returns logical true if the strings are completely identical (which infers that they must be of the same length, also) or logical false if the strings are not the same length or any corresponding characters are not identical. Here are some examples of these comparisons:

PRACTICE 6.4

Assume that these expressions are typed sequentially in the Command Window. Think about it, write down what you think the results will be, and then verify your answers by actually typing them.

```
wxyzstring = ...
'123456789012345';
longstring = '  abc  de f  '
length(longstring)
shortstring =
strtrim(longstring)
length(shortstring)
upper(shortstring)
news = sprintf('The first part is %s', ...
   shortstring(1:3))
```

```
>> word1 = 'cat';
>> word2 = 'car';
>> word3 = 'cathedral';
>> word4 = 'CAR';
>> strcmp(word1,word2)
ans =
     0
>> strcmp(word1,word3)
ans =
     0
>> strcmp(word1,word1)
ans =
     1
>> strcmp(word2,word4)
ans =
     0
```

The function **strncmp** compares only the first n characters in strings and ignores the rest. The first two arguments are the strings to compare, and the third argument is the number of characters to compare (the value of n).

```
>> strncmp(word1,word3,3)
ans =
     1
>> strncmp(word1,word3,4)
ans =
     0
```

QUICK QUESTION!

How can we compare strings, ignoring whether the characters in the string are uppercase or lowercase?

Answer: See the Programming Concept and Efficient Method below.

The Programming Concept

Ignoring the case when comparing strings can be done by changing all characters in the strings to either upper- or lowercase; for example, in MATLAB using the **upper** or **lower** function:

```
>> strcmp(upper(word2),upper(word4))
ans =
     1
```

The Efficient Method

The function **strcmpi** compares the strings but ignores the case of the characters.

```
>> strcmpi(word2,word4)
ans =
     1
```

There is also a function **strncmpi** that compares *n* characters, ignoring the case.

6.2.6 Finding, Replacing, and Separating Strings

There are several functions that find and replace strings, or parts of strings, within other strings and functions that separate strings into substrings.

The function **findstr** receives two strings as input arguments. It finds all occurrences of the shorter string within the longer, and returns the subscripts of the beginning of the occurrences. The order of the strings does not matter with **findstr**; it will always find the shorter string within the longer, whichever that is. The shorter string can consist of one character, or any number of characters. If there is more than one occurrence of the shorter string within the longer one, **findstr** returns a vector with all indices. Note that what is returned is the index of the beginning of the shorter string.

```
>> findstr('abcde', 'd')
ans =
     4
>> findstr('d','abcde')
ans =
     4
>> findstr('abcde', 'bc')
ans =
     2
>> findstr('abcdeabcdedd', 'd')
ans =
     4    9    11    12
```

The function **strfind** does essentially the same thing, except that the order of the arguments does make a difference. The general form is **strfind(string, substring)**; it finds all occurrences of the substring within the string, and returns the subscripts.

```
>> strfind('abcdeabcde','e')
ans =
     5    10
```

For both **strfind** and **findstr**, if there are no occurrences, the empty vector is returned.

```
>> strfind('abcdeabcde','ef')
ans =
     []
```

QUICK QUESTION!

How can you find how many blanks there are in a string (e.g., 'how are you')?

Answer: The **strfind** function will return an index for every occurrence of a substring within a string, so the result is a vector of indices. The **length** of this vector of indices would be the number of occurrences. For example, the following finds the number of blank spaces in *phrase*.

```
>> phrase = 'Hello, and how are you doing?';
>> length(strfind(phrase,' '))
```

```
ans =
    5
```

If you want to get rid of any leading and trailing blanks first (in case there are any), the **strtrim** function would be used first.

```
>> phrase = '   Well, hello there!   ';
>> length(strfind(strtrim(phrase),' '))
ans =
    2
```

Let's expand this, and write a script that creates a vector of strings that are phrases. The output is not suppressed so that the strings can be seen when the script is executed. It loops through this vector and passes each string to a function **countblanks**. This function counts the number of blank spaces in the string, not including any leading or trailing blanks.

phraseblanks.m

```
% This script creates a column vector of phrases
% It loops to call a function to count the number
% of blanks in each one and prints that
phrasemat = char('Hello and how are you?', ...
   'Hi there everyone!', 'How is it going?', 'Whazzup?')
[r c] = size(phrasemat);
for i = 1:r
  % Pass each row (each string) to countblanks function
  howmany = countblanks(phrasemat(i,:));
  fprintf('Phrase %d had %d blanks\n',i,howmany)
end
```

countblanks.m

```
function num = countblanks(phrase)
% Counts the number of blanks in a trimmed string
num = length(strfind(strtrim(phrase), ' '));
```

For example, running this script would result in:

```
>> phraseblanks
phrasemat =
Hello and how are you?
Hi there everyone!
How is it going?
Whazzup?
Phrase 1 had 4 blanks
Phrase 2 had 2 blanks
Phrase 3 had 3 blanks
Phrase 4 had 0 blanks
```

The function **strrep** finds all occurrences of a substring within a string, and replaces them with a new substring. The order of the arguments matters. The format is:

```
strrep(string, oldsubstring, newsubstring)
```

The following example replaces all occurrences of the substring 'e' with the substring 'x':

```
>> strrep('abcdeabcde','e','x')
ans =
abcdxabcdx
```

All strings can be any length, and the lengths of the old and new substrings do not have to be the same.

In addition to the string functions that find and replace, there is a function that separates a string into two substrings. The **strtok** function breaks a string into pieces; it can be called several ways. The function receives one string as an input argument. It looks for the first *delimiter*, which is a character or set of characters that act as a separator within the string. By default, the delimiter is any whitespace character. The function returns a *token*, which is the beginning of the string, up to (but not including) the first delimiter. It also returns the rest of the string, which includes the delimiter. Assigning the returned values to a vector of two variables will capture both of these. The format is

```
[token rest] = strtok(string)
```

where *token* and *rest* are variable names. For example,

```
>> sentence1 = 'Hello there'
sentence1 =
Hello there
>> [word rest] = strtok(sentence1)
word =
Hello
```

```
rest =
there
>> length(word)
ans =
     5
>> length(rest)
ans =
     6
```

Notice that the rest of the string includes the blank space delimiter.

By default, the delimiter for the token is a whitespace character (meaning that the token is defined as everything up to the blank space), but alternate delimiters can be defined. The format

```
[token rest] = strtok(string, delimeters)
```

returns a token that is the beginning of the string, up to the first character contained within the delimiters string, and also the rest of the string. In the following example, the delimiter is the character 'l'.

```
>> [word rest] = strtok(sentence1,'l')
word =
He
rest =
llo there
```

Leading delimiter characters are ignored, whether it is the default whitespace or a specified delimiter. For example, the leading blanks are ignored here:

```
>> [firstpart lastpart] = strtok('   materials science')
firstpart =
materials
lastpart =
 science
```

QUICK QUESTION!

What do you think **strtok** returns if the delimiter is not in the string?

Answer: The first result returned will be the entire string, and the second will be the empty string.

```
>> [first rest] = strtok ('ABCDE')
first =
ABCDE
rest =
  Empty string: 1-by-0
```

QUICK QUESTION!

The function **date** returns the current date as a string; for example, '07-Feb-2008'. How could we write a function to return the day, month, and year as separate output arguments?

Answer: We could use **strrep** to replace the '-' characters with blanks, and then use **strtok** with the blank as the default delimiter to break up the string (twice), or more simply we could just use **strtok** and specify the '-' character as the delimiter.

the delimiter in '-Feb-2008'.) Finally, we need to remove the '-' from the string '-2008'; this can be done by just indexing from the second character to the end of the string.

Here is an example of calling this function:

```
>> [d m y] = separatedate()
d =
07
m =
Feb
y =
2008
```

separatedate.m

```
function [todayday, todaymo, todayyr] = separatedate()
% This function separates the current date into day,
% month, and year
[todayday rest] = strtok(date,'-');
[todaymo todayyr] = strtok(rest,'-');
todayyr = todayyr(2:end);
```

Since we need to separate the string into three parts, we need to use the **strtok** function twice. The first time the string is separated into '07' and '-Feb-2008' using **strtok**. Then, the second string is separated into 'Feb' and '-2008' using **strtok**. (Since leading delimiters are ignored the second '-' is found as

Notice that no input arguments are passed to the function; instead, the **date** function returns the current date as a string.

6.2.7 Evaluating a String

The function **eval** is used to evaluate a string as a function. For example, in the following, the string 'plot(x)' is interpreted to be a call to the **plot** function, and it produces the plot shown in Figure 6.2.

```
>> x = [2 6 8 3];
>> eval('plot(x)')
```

This would be useful if the user entered the name of the type of plot to use. In this example, the string that the user enters (in this case 'bar') is concatenated with the string '(x)' to create the string 'bar(x)'; this is then evaluated as a call to the **bar** function as seen in Figure 6.3. The name of the plot type is also used in the title.

 PRACTICE 6.5

Think about what would be returned by the following sequence of expressions and statements, and then type them into MATLAB to verify your results.

```
strcmp('hello', 'height')
strncmp('hello', 'height',2)
strncmpi('yes', 'YES', 1)
name = 'Smith, Carly';
ind = findstr(name,',')
first = name(1:ind-1)
last = name(ind+2:end)
[f rest] = strtok(name, ',')
l = rest(3:end)
```

FIGURE 6.2

*Plot type passed to the **eval** function.*

FIGURE 6.3

Plot type entered by the user.

 PRACTICE 6.6

Create an *x* vector. Prompt the user for 'sin', 'cos', or 'tan' and create a string with that function of *x* (e.g., 'sin(x)' or 'cos(x)'). Use **eval** to create a *y* vector using the specified function.

```
>> x = [2 6 8 3];
>> whatplot = input('What type of plot?: ', 's');
What type of plot?: bar
>> eval([whatplot '(x)'])
>> title(whatplot)
```

6.3 THE IS FUNCTIONS FOR STRINGS

There are several **is** functions for strings, which return logical true or false. The function **isletter** returns logical true if the character is a letter of the alphabet. The function **isspace** returns logical true if the character is a whitespace character. If strings are passed to these functions, they will return logical true or false for every element, or, in other words, every character.

```
>> isletter('a')
ans =
    1
>> isletter('EK127')
ans =
    1    1    0    0    0
>> isspace('a b')
ans =
    0    1    0
```

The **ischar** function will return logical true if an array is a character array, or logical false if not.

```
>> vec = 'EK127';
>> ischar(vec)
ans =
     1
>> vec = 3:5;
>> ischar(vec)
ans =
     0
```

6.4 CONVERTING BETWEEN STRING AND NUMBER TYPES

MATLAB has several functions that convert numbers to strings in which each character element is a separate digit, and vice versa. (Note: these are different from the functions **char, double**, etc., that convert characters to ASCII equivalents and vice versa.)

To convert numbers to strings, MATLAB has the functions **int2str** for integers and **num2str** for real numbers (which also works with integers). The function **int2str** would convert, for example, the integer 4 to the string '4'.

```
>> rani = randint(1,1,50)
rani =
       38
>> s1 = int2str(rani)
s1 =
38
>> length(rani)
ans =
     1
>> length(s1)
ans =
     2
```

The variable *rani* is a scalar that stores one number, whereas *s1* is a string that stores two characters, '3' and '8'.

Even though the result of the first two assignments is 38, notice that the indentation in the Command Window is different for the number and the string.

The **num2str** function, which converts real numbers, can be called in several ways. If only the real number is passed to the **num2str** function, it will create a string that has four decimal places, which is the default in MATLAB for displaying real numbers. The precision can also be specified (which is the number of digits), and format strings can also be passed, as shown:

```
>> str2 = num2str(3.456789)
str2 =
```

PRACTICE 6.7

Think about what would be returned by the following sequence of expressions and statements, and then type them into MATLAB to verify your results.

```
isletter ('?')
isspace('Oh no!')
str = '12 33';
ischar(str)
v = str2num(str)
ischar(v)
sum(v)
num = 234;
size(num)
snum = int2str(num);
size(snum)
```

```
3.4568
>> length(str2)
ans =
        6
>> str3 = num2str(3.456789,3)
str3 =
3.46
>> str = num2str(3.456789,'%6.2f')
str =
3.46
```

Note that in the last example, MATLAB removed the leading blanks from the string.

The function **str2num** does the reverse; it takes a string in which a number is stored and converts it to the type **double**:

```
>> num = str2num('123.456')
num =
    123.4560
```

If there is a string in which there are numbers separated by blanks, the **str2num** function will convert this to a vector of numbers (of the default type double). For example,

```
>> mystr = '66 2 111';
>> numvec = str2num(mystr)
numvec =
        66      2       111
>> sum(numvec)
ans =
        179
```

QUICK QUESTION!

Let's say that we have a string that consists of an angle followed by either 'd' for degrees or 'r' for radians. For example, it may be a string entered by the user:

```
degrad = input('Enter angle and d/r: ', 's');
    Enter angle and d/r: 54r
```

How could we separate the string into the angle and the character, and then get the sine of that angle using either **sin** or **sind**, as appropriate (**sin** for radians or **sind** for degrees)?

Answer: First, we could separate this string into its two parts:

```
>> angle = degrad(1:length(degrad)-1)
angle =
54
>> dorr = degrad(end)
dorr =
r
```

Then, using an **if-else** statement, we would decide whether to use the **sin** or **sind** function, based on the value of the
(Continued)

variable *dorr*. Let's assume that the value is 'r' so we want to use **sin**. The variable *angle* is a string so the following would not work:

```
>> sin(angle)
??? Function 'sin' is not defined
for values of class 'char'.
>>
```

Instead, we could use either **str2num** to convert the string to a number, or concatenation to create a string 'sin(54)' (or

whatever the value of the variable *angle* is) and pass that to the **eval** function:

```
>> eval(['sin(' angle ')'])
ans =
  −0.5588
>> sin(str2num(angle))
ans =
  −0.5588
```

A complete script to accomplish this is shown here.

angle_d_or_r.m

```
% Prompt the user for angle and 'd' for degrees
% or 'r' for radians; print the sine of the angle

% Read in the response as a string and then
% separate the angle and character
degrad = input('Enter angle and d/r: ', 's');
angle = degrad(1:length(degrad)-1);
dorr = degrad(end);
% Error-check to make sure user enters 'd' or 'r'
while dorr ~= 'd' & dorr ~= 'r'
    disp('Error! Enter d or r with the angle.')
    degrad = input('Enter angle and d/r: ', 's');
    angle = degrad(1:length(degrad)-1);
    dorr = degrad(end);
end
% Convert angle to number
anglenum = str2num(angle);
fprintf('The sine of %.1f ', anglenum)
% Choose sin or sind function
if dorr == 'd'
    fprintf('degrees is %.3f.\n', sind(anglenum))
else
    fprintf('radians is %.3f.\n', sin(anglenum))
end
```

(Continued)

```
>> angle_d_or_r
Enter angle and d/r: 3.1r
The sine of 3.1 radians is 0.042.
>> angle_d_or_r
Enter angle and d/r: 53t
Error! Enter d or r with the angle.
Enter angle and d/r: 53d
The sine of 53.0 degrees is 0.799.
```

SUMMARY

Common Pitfalls

- Putting arguments to **strfind** in incorrect order (the order matters for **strfind** but not for **findstr**).

- Confusing **sprintf** and **fprintf**. The syntax is the same, but **sprintf** creates a string whereas **fprintf** prints.

- Trying to create a vector of strings with varying lengths (the easiest way is to use **strvcat** or **char**, which will pad with extra blanks automatically).

- Forgetting that when using **strtok**, the second argument returned (the rest of the string) contains the delimiter.

- When breaking a string into pieces, forgetting to convert the numbers in the strings to actual numbers that can then be used in calculations.

Programming Style Guidelines

- Trim trailing blanks from strings stored in matrices before using.

- Make sure the correct string comparison function is used; for example, **strcmpi** if ignoring case is desired.

MATLAB Functions and Commands			
strcat	upper	findstr	isletter
strvcat	lower	strfind	isspace
blanks	strcmp	strtok	ischar
sprintf	strncmp	strrep	int2str
deblank	strcmpi	date	num2str
strtrim	strncmpi	eval	str2num

Exercises

1. Write a script that will prompt the user to enter a word, and then print the first character in the word. For example, the output might look like this:

```
>> Enter a word: howdy
The word howdy starts with the letter 'h'
```

2. Write a function that will receive a name and department as separate strings and will create and return a code consisting of the first two letters of the name and the last two letters of the department. The code should be uppercase letters. For example,

```
>> namedept('Robert','Mechanical')
ans =
ROAL
```

3. Write a function that will prompt the user separately for a first and last name and will create and return a string with the form 'last, first'.

4. Write a function that will prompt the user separately for a filename and extension and will create and return a string with the form 'filename.ext'.

5. Write a script that will, in a loop, prompt the user for four course numbers. Each will be a string of length 5 of the form 'CS101'. These strings are to be stored in a character matrix.

6. Write a function that will receive one input argument, which is an integer n. The function will prompt the user for a number in the range from 1 to n (the actual value of n should be printed in the prompt) and return the user's input. The function should error-check to make sure that the user's input is in the correct range.

7. Write a script that will create x and y vectors. Then, it will ask the user for a color ('red', 'blue', or 'green') and for a plot style ('o', '*'). It will then create a string *pstr* that contains the color and plot style, so that the call to the **plot** function would be **plot(x,y,pstr)**. For example, if the user enters 'blue' and '*', the variable pstr would contain 'b*'.

8. Write a script that will generate a random integer, ask the user for a field width, and print the random integer with the specified field width. The script will use **sprintf** to create a string such as 'The # is %4d\n' (if, for example, the user entered 4 for the field width), which is then passed to the **fprintf** function. In order to print (or create a string using **sprintf**) either the '%' or '\' character, there must be two of them in a row.

9. What does the **blanks** function return when a 0 is passed to it? A negative number? Write a function *myblanks* that does exactly the same thing as the **blanks** function. Here are some examples of calling it:

```
>> fprintf('Here is the result:%s!\n', myblanks(0))
Here is the result:!
>> fprintf('Here is the result:%s!\n', myblanks(7))
Here is the result:       !
```

10. Write a function that will receive two strings as input arguments, and will return a character matrix with the two strings in separate rows. Rather than using the

char function to accomplish this, the function should pad with extra blanks as necessary and create the matrix using square brackets.

11. The functions that label the x and y axes and title on a plot expect string arguments. These arguments can be string variables. Write a script that will prompt the user for an integer n, then create an x vector with integer values from 1 to n, a y vector that is x^2, and then plot with a title that says x^2 with n values where the number is actually in the title.

12. Load files named 'file1.dat', 'file2.dat', and so on in a loop. To test this, create just two files with these names in your current directory first.

13. Write a script *trimblanks* that will prompt the user for a string, and then remove the leading and trailing blanks. It will print the string, with quotes, both with and without the leading and trailing blanks. Here is an example of calling it:

```
>> trimblanks
Enter a string:     abcde  fg
You entered '     abcde  fg '
Without blanks, it is 'abcde  fg'
```

14. Write a function that will receive two input arguments: a character matrix that is a column vector of strings, and a string. It will loop to look for the string within the character matrix. The function will return the row number in which the string is found if it is in the character matrix, or the empty vector if not.

15. If the strings passed to **strfind** or **findstr** are the same length, what are the only two possible results that could be returned?

16. Either in a script or in the Command Window, create a string variable that stores a string in which numbers are separated by the character 'x', for example '12x3x45x2'. Create a vector of the numbers, and then get the sum (e.g., for the example given it would be 62 but the solution should be general).

17. Assembly language instructions frequently are in the form of a word that represents the operator and then the operands separated by a comma. For example, the string 'ADD n,m' is an instruction to add n + m. Write a function *assembly_add* that will receive a string in this form and will return the sum of n + m. For example,

```
>> assembly_add('ADD 10,11')
ans =
   21
```

18. In cryptography, the intended message sometimes consists of the first letter of every word in a string. Write a function *crypt* that will receive a string with the encrypted message and return the message.

```
>> estring = 'The early songbird tweets';
>> m = crypt(estring)
m =
Test
```

19. Write a function *rid_multiple_blanks* that will receive a string as an input argument. The string contains a sentence that has multiple blank spaces in between some of the words. The function will return the string with only one blank in between words. For example,

```
>> mystr = 'Hello    and how   are    you?';
>> rid_multiple_blanks(mystr)
ans =
Hello and how are you?
```

20. Words in a sentence variable (just a string variable) called *mysent* are separated by /'s instead of blank spaces. For example, mysent might have this value:

```
'This/is/not/quite/right'
```

Write a function *slashtoblank* that will receive a string in this form and will return a string in which the words are separated by blank spaces. This should be general and work regardless of the value of the argument. No loops are allowed in this function; the built-in string function(s) must be used.

```
>> mysent = 'This/is/not/quite/right';
>> newsent = slashtoblank(mysent)
newsent =
This is not quite right
```

21. Create the following two variables:

```
>> var1 = 123;
>> var2 = '123';
```

Then, add 1 to each of the variables. What is the difference?

22. A filename is supposed to be in the form 'filename.ext'. Write a function that will determine whether a string is in the form of a name followed by a dot followed by a three-character extension, or not. The function should return 1 for logical true if it is in that form, or 0 for false if not.

23. The built-in **clock** function returns a vector with six elements representing the year, month, day, hours, minutes, and seconds; the first five elements are integers whereas the last is a **double** value, but calling it with **fix** will convert all to integers. The built-in **date** function returns the day, month, and year as a string. For example,

```
>> fix(clock)
ans =
   2008    4   25   14   25   49
>> date
ans =
25-Apr-2008
```

Write a script that will call both of these built-in functions, and then compare results to make sure that the year is the same. The script will have to convert one from a string to a number, or the other from a number to a string in order to compare.

24. Find out how to pass a vector of integers to **int2str** or real numbers to **num2str**.

25. Write a function *wordscramble* that will receive a word in a string as an input argument. It will then randomly scramble the letters and return the result. Here are examples of calling the function:

```
>> wordscramble('fantastic')
ans =
safntcait
>> sc = wordscramble('hello')
sc =
hleol
```

26. Using the functions **char** and **double**, you can shift words. For example, you can convert from lowercase to uppercase by subtracting 32 from the character codes:

```
>> orig = 'ape';
>> new = char(double(orig)-32)
new =
APE
>> char(double(new)+32)
ans =
ape
```

We've encrypted a string by altering the character codes. Figure out the original string. Try adding and subtracting different values (do this in a loop) until you decipher it:

```
Jmkyvih$mx$syx$}ixC
```

27. Two variables store strings that consist of a letter of the alphabet, a blank space, and a number (in the form 'r 14.3'). Write a script that would initialize two such variables. Then, use string manipulating functions to extract the numbers from the strings and add them together.

28. Write a script that will first initialize a string variable that will store x and y coordinates of a point in the form 'x 3.1 y 6.4'. Then, use string manipulating functions to extract the coordinates and plot them.

29. Modify the script in the previous example to be more general: the string could store the coordinates in any order; for example, it could store 'y 6.4 x 3.1'.

30. Write a script that will be a temperature converter. The script prompts the user for a temperature in degrees Fahrenheit, and then uses the **menu** function to let the user choose whether to convert that temperature to degrees Celsius or degrees Kelvin. The user's temperature should be in the title of the menu. The script will then print the converted temperature. The conversions are C = (F − 32) * 5/9 and K = C + 273.15.

Data Structures: Cell Arrays and Structures

Data structures are variables that store more than one value. In order for it to make sense to store more than one value in a variable, the values should somehow be logically related. There are many different kinds of data structures. We have already been working with one kind, arrays (e.g., vectors and matrices). An array is a data structure in which all the values are logically related in that they are of the same type, and represent in some sense the same thing. So far, that has been true for the vectors and matrices that we have used.

A *cell array* is a kind of data structure that stores values of different types. Cell arrays can be vectors or matrices; the different values are stored in the elements of the array. One very common use of a cell array is to store strings of different lengths.

Structures are data structures that group together values that are logically related, but are not the same thing and not necessarily the same type. The different values are stored in separate *fields* of the structure.

One use of structures is to set up a *database* of information. For example, for a class a professor might want to store information for every student in the class: the student's name, university ID number, grades on all assignments and quizzes, and so on. In many programming languages and database programs, the terminology is that within a database file, there would be one ***record*** of information for each student; each separate piece of information (name, quiz 1 score, etc.) would be called a ***field*** of the record. In the MATLAB® software these records are called ***structs***.

7.1 CELL ARRAYS

One type of data structure that MATLAB has but is not found in many programming languages is a ***cell array***. A cell array in MATLAB is an array, but unlike the vectors and matrices we have used so far, elements in cell arrays can store different types of values.

7.1.1 Creating Cell Arrays

There are several ways to create cell arrays. For example, we will create a cell array in which one element will store an integer, one element will store a character, one element will store a vector, and one element will store a string. Just as with the arrays we have seen so far, this could be a 1×4 row vector, a 4×1 column vector, or a 2×2 matrix. The syntax for creating vectors and matrices is the same as before. Values within rows are separated by spaces or commas, and rows are separated by semicolons. However, for cell arrays, curly braces are used rather than square brackets. For example, the following creates a row vector cell array with the four different types of values:

```
>> cellrowvec = {23, 'a', 1:2:9, 'hello'}
cellrowvec =
    [23]    'a'    [1x5 double]    'hello'
```

To create a column vector cell array, the values are instead separated by semicolons:

```
>> cellcolvec = {23; 'a'; 1:2:9; 'hello'}
cellcolvec =
    [       23]
    'a'
    [1x5 double]
    'hello'
```

This method creates a 2×2 cell array matrix:

```
>> cellmat = {23 'a'; 1:2:9 'hello'}
cellmat =
    [       23]    'a'
    [1x5 double]    'hello'
```

Another method of creating a cell array is simply to assign values to specific array elements and build it up element by element. However, as explained before, extending an array element by element is a very inefficient and time-consuming

method. It is much more efficient, if the size is known ahead of time, to preallocate the array. For cell arrays, this is done with the **cell** function. For example, to
preallocate a variable *mycellmat* to be a 2 × 2 cell array, the **cell** function would
be called as follows:

```
>> mycellmat = cell(2,2)
mycellmat =
      []    []
      []    []
```

Note that this is a function call so the arguments to the function are in parentheses. This creates a matrix in which all the elements are empty vectors. Then,
each element can be replaced by the desired value. How to refer to each element in order to accomplish this will be explained next.

7.1.1.1 Referring to and Displaying Cell Array Elements and Attributes

Just as with the other vectors we have seen so far, we can refer to individual elements of cell arrays. The only difference is that curly braces are used for the subscripts. For example, this refers to the second element of the cell array *cellrowvec*:

```
> cellrowvec{2}
ans =
a
```

Row and column indices are used to refer to an element in a matrix (again
using curly braces), for example,

```
>> cellmat{1,1}
ans =
      23
```

Values can be assigned to cell array elements. For example, after preallocating
the variable *mycellmat* in the previous section, the elements can be initialized:

```
>> mycellmat{1,1} = 23
mycellmat =
      [23]   []
       []    []
```

When an element of a cell array is itself a data structure, only the type of the
element is displayed when the cell array contents are shown. For example, in
the cell arrays just created, the vector is shown just as *1 × 5* **double**. Referring to
that element specifically would display its contents, for example,

```
>> cellmat{2,1}
ans =
      1   3   5   7   9
```

Since this element is a vector, parentheses are used to refer to its elements. For example, the fourth element of the preceding vector is:

```
>> cellmat{2,1}(4)
ans =
     7
```

Notice that the index into the cell array is given in curly braces, and then parentheses are used to refer to an element of the vector.

We can also refer to subsets of cell arrays, for example,

```
>> cellcolvec{2:3}
ans =
a
ans =
     1   3   5   7   9
```

Notice, however, that MATLAB stored *cellcolvec*{2} in the default variable *ans*, and then replaced that with the value of *cellcolvec*{3}. This is because the two values are different types, and therefore cannot be stored together in *ans*. However, they could be stored in two separate variables by having a vector of variables on the left-hand side of an assignment.

```
>> [c1 c2] = cellcolvec{2:3}
c1 =
a
c2 =
     1   3   5   7   9
```

There are several methods of displaying cell arrays. The **celldisp** function displays all elements of the cell array:

```
>> celldisp(cellrowvec)
cellrowvec{1} =
       23
cellrowvec{2} =
a
cellrowvec{3} =
     1   3   5   7   9
cellrowvec{4} =
hello
```

The function **cellplot** puts a graphical display of the cell array in a Figure Window; however, it is a high-level view and basically just displays the same information as typing the name of the variable (e.g., it wouldn't show the contents of the vector in the previous example).

Many of the functions and operations on arrays that we have already seen also work with cell arrays. For example, here are some related to dimensioning:

```
>> length(cellrowvec)
ans =
       4
>> size(cellcolvec)
ans =
       4    1
>> cellrowvec{end}
ans =
hello
```

It is not possible to delete an individual element from a cell array. For example, assigning an empty vector to a cell array element does not delete the element, it just replaces it with the empty vector:

```
>> cellrowvec
mycell =
       [23]    'a'      [1x5 double]     'hello'
>> length(cellrowvec)
ans =
       4
>> cellrowvec{2} = []
mycell =
       [23]     []      [1x5 double]     'hello'
>> length(cellrowvec)
ans =
       4
```

However, it is possible to delete an entire row or column from a cell array by assigning the empty vector (**Note**: use parentheses rather than curly braces to refer to the row or column):

```
>> cellmat
mycellmat =
       [      23]        'a'
       [1x5 double]     'hello'
>> cellmat(1,:) = []
mycellmat =
       [1x5 double]     'hello'
```

7.1.1.2 Storing Strings in Cell Arrays
One good application of a cell array is to store strings of different lengths. Since cell arrays can store different types of values in the elements, that means strings of different lengths can be stored in the elements.

```
>> names = {'Sue', 'Cathy', 'Xavier'}
names =
       'Sue'    'Cathy'    'Xavier'
```

This is extremely useful, because unlike vectors of strings created using **char** or **strvcat**, these strings do not have extra trailing blanks.

The length of each string can be displayed using a **for** loop to loop through the elements of the cell array:

```
>> for i = 1:length(names)
     disp(length(names{i}))
   end
   3
   5
   6
```

It is possible to convert from a cell array of strings to a character array, and vice versa. MATLAB has several functions that facilitate this. For example, the function **cellstr** converts from a character array padded with blanks to a cell array in which the trailing blanks have been removed.

```
>> greetmat = char('Hello','Goodbye');
>> cellgreets = cellstr(greetmat)
cellgreets =
     'Hello'
     'Goodbye'
```

The **char** function can convert from a cell array to a character matrix:

```
>> names = {'Sue', 'Cathy', 'Xavier'};
>> cnames = char(names)
cnames =
Sue
Cathy
Xavier
>> size(cnames)
ans =
     3     6
```

The function **iscellstr** will return logical true if a cell array is a cell array of all strings, or logical false if not.

```
>> iscellstr(names)
ans =
     1
>> iscellstr(cellcolvec)
ans =
     0
```

We will see several examples of cell arrays containing strings of varying lengths in the coming chapters, including advanced file input functions and customizing plots.

7.2 STRUCTURES

Structures are data structures that group together values that are logically related in what are called *fields* of the structure. An advantage of structures is that the fields are named, which helps to make it clear what the values are that are stored in the structure. However, structure variables are not arrays. They do not have elements, so it is not possible to loop through the values in a structure.

7.2.1 Creating and Modifying Structure Variables

Creating structure variables can be accomplished by simply storing values in fields using assignment statements, or by using the **struct** function.

The first example that will be used is that the local Computer Super Mart wants to store information on the software packages that it sells. For every one, they will store:

- The item number

- The cost to the store

- The price to the customer

- A code indicating the type of software

An individual structure variable for one software package might look like this:

package			
item_no	**cost**	**price**	**code**
123	19.99	39.95	'g'

The name of the structure variable is *package*; it has four fields called item_no, cost, price, and code.

One way to initialize a structure variable is to use the **struct** function, which preallocates the structure. The field names are passed to the struct in quotes, following each one with the value for that field:

```
>> package = struct('item_no',123,'cost',19.99,...
     'price',39.95,'code','g')
package =
    item_no: 123
       cost: 19.9900
      price: 39.9500
       code: 'g'
```

Typing the name of the structure variable will display the names and contents of all fields:

```
>> package
package =
    item_no: 123
       cost: 19.9900
      price: 39.9500
       code: 'g'
```

Notice that in the Workspace Window, the variable *package* is listed as a 1×1 struct. MATLAB, since it is written to work with arrays, assumes the array format. Just as a single number is treated as a 1×1 double, a single structure is treated as a 1×1 struct. Later in this chapter we will see how to work more generally with vectors of structs.

An alternative method of creating this structure, which is not as efficient, involves using the ***dot operator*** to refer to fields within the structure. The name of the structure variable is followed by a dot, or period, and then the name of the field within that structure. Assignment statements can be used to assign values to the fields.

```
>> package.item_no = 123;
>> package.cost = 19.99;
>> package.price = 39.95;
>> package.code = 'g';
```

By using the dot operator in the first assignment statement, a structure variable is created with the field *item_no*. The next three assignment statements add more fields to the structure variable.

Adding a field to a structure later is accomplished as shown earlier, by using an assignment statement.

An entire structure variable can be assigned to another. This would make sense, for example, if the two structures had some values in common. Here, for example, the values from one structure are copied into another and then two fields are selectively changed.

```
>> newpack = package;
>> newpack.item_no = 111;
>> newpack.price = 34.95
newpack =
    item_no: 111
       cost: 19.9900
      price: 34.9500
       code: 'g'
```

To print from a structure, the **disp** function will display either the entire structure or a field.

```
>> disp(package)
     item_no: 123
        cost: 19.9900
       price: 39.9500
        code: 'g'
>> disp(package.cost)
    19.9900
```

However, using **fprintf**, only individual fields can be printed; the entire structure cannot be printed.

```
>> fprintf('%d %c\n', package.item_no, package.code)
123 g
```

The function **rmfield** removes a field from a structure. It returns a new structure with the field removed, but does not modify the original structure (unless the returned structure is assigned to that variable). For example, the following would remove the *code* field from the *newpack* structure, but store the resulting structure in the default variable *ans*. The value of *newpack* remains unchanged.

```
>> rmfield(newpack, 'code')
ans =
     item_no: 111
        cost: 19.9900
       price: 34.9500
>> newpack
newpack =
     item_no: 111
        cost: 19.9000
       price: 34.9500
        code: 'g'
```

To change the value of *newpack*, the structure that results from calling **rmfield** must be assigned to *newpack*.

```
>> newpack = rmfield(newpack, 'code')
newpack =
     item_no: 111
        cost: 19.9000
       price: 34.9500
```

7.2.2 Passing Structures to Functions

An entire structure can be passed to a function, or individual fields can be passed. For example, here are

 PRACTICE 7.2

A silicon wafer manufacturer stores, for every part in their inventory, a part number, how many are in the factory, and the cost for each.

onepart		
part_no	**quantity**	**costper**
123	4	33

Create this structure variable using **struct**. Print the cost in the form $xx.xx.

two different versions of a function that calculates the profit on a software package. The profit is defined as the price minus the cost.

In the first version, the entire structure variable is passed to the function, so the function must use the dot operator to refer to the *price* and *cost* fields of the input argument.

calcprof.m
```
function profit = calcprof(packstruct)
% Calculates the profit for a software package
% The entire structure is passed to the function
profit = packstruct.price - packstruct.cost;
```

```
>> calcprof(package)
ans =
    19.9600
```

In the second version, just the *price* and *cost* fields are passed to the function using the dot operator in the function call. These are passed to two scalar input arguments in the function header, so there is no reference to a structure variable in the function itself, and the dot operator is not needed in the function.

calcprof2.m
```
function profit = calcprof2(oneprice, onecost)
% Calculates the profit for a software package
% The individual fields are passed to the function
profit = oneprice - onecost;
```

```
>> calcprof2(package.price, package.cost)
ans =
    19.9600
```

It is important, as always with functions, to make sure that the arguments in the function call correspond one-to-one with the input arguments in the function header. In the case of *calcprof*, a structure variable is passed to an input argument, which is a structure. For the second function *calcprof2*, two individual fields, which are **double** values, are passed to two **double** arguments.

7.2.3 Related Structure Functions

There are several functions that can be used with structures in MATLAB. The function **isstruct** will return 1 for logical true if the variable argument is a structure variable, or 0 if not. The **isfield** function returns logical true if a fieldname (as a string) is a field in the structure argument, or logical false if not.

```
>> isstruct(package)
ans =
     1
>> isfield(package,'cost')
ans =
     1
```

The **fieldnames** function will return the names of the fields that are contained in a structure variable.

```
>> pack_fields = fieldnames(package)
pack_fields =
     'item_no'
     'cost'
     'price'
     'code'
```

Since the names of the fields are of varying lengths, the **fieldnames** function returns a cell array with the names of the fields.

Curly braces are used to refer to the elements, since pack_fields is a cell array. For example, we can refer to the length of one of the strings:

```
>> length(pack_fields{2})
ans =
     4
```

QUICK QUESTION!

How can we ask the user for a field in a structure and either print its value or an error if it is not actually a field?
Answer: The **isfield** function can be used to determine whether or not it is a field of the structure. Then, by concat-enating that field name to the structure variable and dot, and then passing the entire string to **eval**, the expression would be evaluated to the actual field in the structure. The following code

```
inputfield = input('Which field would you like to see: ','s');
if isfield(package, inputfield)
   fprintf('The value of the %s field is: %s\n', ...
      inputfield, eval(['package.' inputfield]))
else
   fprintf('Error: %s is not a valid field\n', inputfield)
end
```

would produce this output (assuming the *package* variable was initialized as shown earlier):

```
Which field would you like to see: code
The value of the code field is: g
```

7.2.4 Vectors of Structures

In many applications, including database applications, information normally would be stored in a *vector of structures*, rather than in individual structure variables. For example, if the Computer Super Mart is storing information on all the software packages that it sells, it would likely be in a vector of structures, for example,

		packages		
	item_no	**cost**	**price**	**code**
1	123	19.99	39.95	'g'
2	456	5.99	49.99	'l'
3	587	11.11	33.33	'w'

In this example, *packages* is a vector that has three elements. It is shown as a column vector. Each element is a structure consisting of four fields, item_no, cost, price, and code. It may look like a matrix with rows and columns, but it is instead a vector of structures.

This can be created several ways. One method is to create a structure variable, as shown earlier, to store information on one software package. This can then be expanded to be a vector of structures.

```
>> packages = struct('item_no',123,'cost',19.99,...
        'price',39.95,'code','g');
>> packages(2) = struct('item_no',456,'cost', 5.99,...
        'price',49.99,'code','l');
>> packages(3) = struct('item_no',587,'cost',11.11,...
        'price',33.33,'code','w');
```

The first assignment statement shown here creates the first structure in the structure vector, the next one creates the second structure, and so on. This actually creates a 1 × 3 row vector.

Alternatively, the first structure could be treated as a vector to begin with, for example,

```
>> packages(1) = struct('item_no',123,'cost',19.99,...
        'price',39.95,'code','g');
>> packages(2) = struct('item_no',456,'cost', 5.99,...
        'price',49.99,'code','l');
>> packages(3) = struct('item_no',587,'cost',11.11,...
        'price',33.33,'code','w');
```

Both of these methods, however, involve extending the vector. As we have already seen, preallocating any vector in MATLAB is more efficient than extending it. There are several methods of preallocating the vector. By starting with the last element, MATLAB would create a vector with that many elements. Then, the elements from 1 through end-1 could be initialized. For example, for a vector of structures that has three elements, start with the third element.

```
>> packages(3) = struct('item_no',587,'cost',11.11,...
      'price',33.33,'code','w');
>> packages(1) = struct('item_no',123,'cost',19.99,...
      'price',39.95,'code','g');
>> packages(2) = struct('item_no',456,'cost', 5.99,...
      'price',49.99,'code','l');
```

Another method is to create one element with the values from one structure, and use **repmat** to replicate it to the desired size. Then, the remaining elements can be modified. The following creates one structure and then replicates this into a 1×3 matrix.

```
>> packages = repmat(struct('item_no',587,'cost',...
      11.11, 'price',33.33,'code','w'), 1,3);
>> packages(2) = struct('item_no',456,'cost', 5.99,...
      'price',49.99,'code','l');
>> packages(3) = struct('item_no',587,'cost',11.11,...
      'price',33.33,'code','w');
```

Typing the name of the variable will display only the size of the structure vector and the names of the fields:

```
>> packages
packages =
1x3 struct array with fields:
    item_no
    cost
    price
    code
```

The variable *packages* is now a vector of structures, so each element in the vector is a structure. To display one element in the vector (one structure), an index into the vector would be specified. For example, to refer to the second element:

```
>> packages(2)
ans =
    item_no: 456
       cost: 5.9900
      price: 49.9900
       code: 'l'
```

To refer to a field, it is necessary to refer to the particular structure, and then the field within it. This means using an index into the vector to refer to the structure, and then the dot operator to refer to a field. For example:

```
>> packages(1).code
ans =
g
```

So, there are essentially three levels to this data structure. The variable *packages* is the highest level, which is a vector of structures. Each of its elements is an individual structure. The fields within these individual structures are the lowest level. The following loop displays each element in the *packages* vector.

```
>> for i = 1:length(packages)
       disp(packages(i))
   end
   item_no: 123
      cost: 19.9900
     price: 39.9500
      code: 'g'

   item_no: 456
      cost: 5.9900
     price: 49.9900
      code: 'l'

   item_no: 587
      cost: 11.1100
     price: 33.3300
      code: 'w'
```

To refer to a particular field for all structures, in most programming languages it would be necessary to loop through all elements in the vector and use the dot operator to refer to the field for each element. However, this is not the case in MATLAB.

The Programming Concept

For example, to print all of the costs, a **for** loop could be used:

```
>> for i=1:3
       fprintf('%f\n',packages(i).cost)
   end
19.990000
5.990000
11.110000
```

The Efficient Method

However, **fprintf** would do this automatically in MATLAB:

```
>> fprintf('%f\n',packages.cost)
19.990000
5.990000
11.110000
```

Using the dot operator in this manner to refer to all values of a field would result in the values being stored successively in the default variable *ans*:

```
>> packages.cost
ans =
    19.9900
ans =
    5.9900
ans =
    11.1100
```

However, the values can all be stored in a vector:

```
>> pc = [packages.cost]
pc =
    19.9900     5.9900     11.1100
```

Using this method, MATLAB allows the use of functions on all the same fields within a vector of structures. For example, to sum all three cost fields, the vector of cost fields is passed to the **sum** function:

```
>> sum([packages.cost])
ans =
    37.0900
```

For vectors of structures, the entire vector (e.g., *packages*) could be passed to a function, or just one element (e.g., *packages(1)*), which would be a structure, or a field within one of the structures (e.g., *packages(2).price*).

Here is an example of a function that receives the entire vector of structures as an input argument, and prints all of it in a nice table format.

printpackages.m

```
function printpackages(packstruct)
% This function prints a table showing all
% values from a vector of packages structures
fprintf('\nItem # Cost Price Code\n\n')
no_packs = length(packstruct);
for i = 1:no_packs
    fprintf('%6d %6.2f %6.2f %3c\n', ...
        packstruct(i).item_no, ...
        packstruct(i).cost, ...
        packstruct(i).price, ...
        packstruct(i).code)
end
```

The function loops through all the elements of the vector, each of which is a structure, and uses the dot operator to refer to and print each field. Here is an example of calling the function:

```
>> printpackages(packages)
Item #        Cost        Price       Code
123           19.99       39.95       g
456           5.99        49.99       l
587           11.11       33.33       w
```

 PRACTICE 7.3

A silicon wafer manufacturer stores, for every part in their inventory, a part number, how many are in the factory, and the cost for each. First, create a vector of structs called *parts* so that when displayed it has the following values:

```
>> parts
parts =
1x3 struct array with fields:
    partno
    quantity
    costper
>> parts(1)
ans =
      partno: 123
    quantity: 4
     costper: 33
>> parts(2)
ans =
      partno: 142
    quantity: 1
     costper: 150
>> parts(3)
ans =
      partno: 106
    quantity: 20
     costper: 7.5000
```

Next, write general code that will, for any values and any number of structures in the variable *parts*, print the part number and the total cost (quantity of parts multiplied by the cost of each) in a column format. For example, if the variable *parts* stores the values shown, the result would be:

```
123    132.00
142    150.00
106    150.00
```

The previous example involved a vector of structs. In the next example, a some-what more complicated data structure will be introduced: a vector of structs in which some fields are vectors themselves. The example is a database of infor-mation that a professor might store for his or her class. This will be imple-mented as a vector of structures. The vector will store all the class information. Every element in the vector will be a structure, representing all information about one particular student. For every student, the professor wants to store (for now, this would be expanded later):

- Name (a string)

- University ID number

- Quiz scores (a vector of 4 quiz scores)

The vector variable, called *student*, might look like this:

	name	id_no	quiz			
			1	2	3	4
1	C, Joe	999	10.0	9.5	0.0	10.0
2	Hernandez, Pete	784	10.0	10.0	9.0	10.0
3	Brownnose, Violet	332	7.5	6.0	8.5	7.5

Each element in the vector is a struct with three fields (*name, id_no, quiz*). The *quiz* field is a vector of quiz grades. The *name* field is a string.

This data structure could be defined as follows.

```
>> student(3) = struct('name','Brownnose, Violet',...
   'id_no',332,'quiz', [7.5 6 8.5 7.5]);
>> student(1) = struct('name','C, Joe',...
   'id_no',999,'quiz', [10 9.5 0 10]);
>> student(2) = struct('name','Hernandez, Pete',...
   'id_no',784,'quiz', [10 10 9 10]);
```

Once this data structure has been initialized, in MATLAB we could refer to dif-ferent parts of it. The variable *student* is the entire array; MATLAB just shows the names of the fields.

```
>> student
student =
1x3 struct array with fields:
    name
    id_no
    quiz
```

To see the actual values, we would have to refer to individual structures and fields.

```
>> student(1)
ans =
     name: 'C, Joe'
     id_no: 999
     quiz: [10 9.5000 0 10]
>> student(1).quiz
ans =
    10.0000      9.5000      0    10.0000
>> student(1).quiz(2)
ans =
     9.5000
>> student(3).name(1)
ans =
B
```

With a more complicated data structure like this, it is important to be able to understand different parts of the variable. The following are examples of expressions that refer to different parts of this data structure:

- *student* is the entire data structure, which is a vector of structs

- *student(1)* is an element from the vector, which is an individual **struct**

- *student(1).id_no* is the *id_no* field from the structure, which is a **double** value

- *student(1).quiz* is the *quiz* field from the structure, which is a vector of doubles

- *student(1).quiz(1)* is an individual **double** quiz grade

One example of using this data structure would be to calculate and print the quiz average for each student. The following function accomplishes this. The *student* structure, as just defined, is passed to this function. The algorithm in the function is:

- Print column headings.

- Loop through the individual students. For each,

  - Sum the quiz grades

  - Calculate the average

  - Print the student's name and quiz average

With the programming method, a second (nested) loop would be required to find the running sum of the quiz grades. However, as we have seen, the **sum** function can be used to sum the vector of all quiz grades for each student. The function is defined as:

print_aves.m

```
function print_aves(student)
% This function prints the average quiz grade
% for each student in the vector of structs
fprintf('%-20s %-10s\n', 'Name', 'Average')
for i = 1:length(student)
    qsum = sum([student(i).quiz]);
    no_quizzes = length(student(i).quiz);
    ave = qsum / no_quizzes;
    fprintf('%-20s %.1f\n', student(i).name, ave);
end
```

Here is an example of calling the function:

```
>> print_aves(student)
Name                Average
C, Joe              7.4
Hernandez, Pete     9.8
Brownnose, Violet   7.4
```

7.2.5 Nested Structures

A *nested structure* is a structure in which at least one member is itself a structure.

For example, a structure for a line segment might consist of fields representing the two points at the ends of the line segment. Each of these points would be represented as a structure consisting of the x- and y-coordinates.

lineseg			
endpoint1		endpoint2	
x	y	x	y
2	4	1	6

This shows a structure variable called *lineseg* that has two fields, *endpoint1* and *endpoint2*. Each of these is a structure consisting of two fields, *x* and *y*.

One method of defining this is to nest calls to the **struct** function:

```
>> lineseg = struct('endpoint1',struct('x',2,'y',4), ...
                    'endpoint2',struct('x',1,'y',6))
```

This method is the most efficient. However, another method is to build the nested structure one field at a time. Since this is a nested structure with one structure inside of another, the dot operator must be used twice here to get to the actual x- and y-coordinates.

```
>> lineseg.endpoint1.x = 2;
>> lineseg.endpoint1.y = 4;
>> lineseg.endpoint2.x = 1;
>> lineseg.endpoint2.y = 6;
```

Once the nested structure has been created, we can refer to different parts of the variable *lineseg*. Just typing the name of the variable shows only that it is a structure consisting of two fields, *endpoint1* and *endpoint2*, each of which is a structure.

```
>> lineseg
lineseg =
   endpoint1: [1x1 struct]
   endpoint2: [1x1 struct]
```

Typing the name of one of the nested structures will display the field names and values within that structure:

```
>> lineseg.endpoint1
ans =
   x: 2
   y: 4
```

Using the dot operator twice will refer to an individual coordinate, for example,

```
>> lineseg.endpoint1.x
ans =
     2
```

QUICK QUESTION!

How could we write a function *strpoint* that returns a string (x,y) containing the x- and y-coordinates? For example, it might be called separately to create strings for the two end-points and then printed as shown here:

```
>> fprintf('The line segment consists of %s and %s\n', ...
   strpoint(lineseg.endpoint1), ...
   strpoint(lineseg.endpoint2))
The line segment consists of (2, 4) and (1, 6)
```

Answer: Since an endpoint structure is passed to the function to an input argument, the dot operator is used within the function to refer to the x- and y-coordinates. The **sprintf** function is used to create the string that is returned.

(Continued)

```
strpoint.m
function ptstr = strpoint(ptstruct)
% This function receives the struct containing x and y
% coordinates and returns a string '(x,y)'
ptstr = sprintf('(%d, %d)', ptstruct.x, ptstruct.y);
```

A nested structure variable for a line segment could also be created by creating structure variables for the points first, and then storing these in the two fields of a line segment variable. For example:

```
>> pointone = struct('x', 5, 'y', 11);
>> pointtwo = struct('x', 7, 'y', 9);
>> myline = struct('endpoint1', pointone,...
                   'endpoint2', pointtwo)
myline =
  endpoint1: [1x1 struct]
  endpoint2: [1x1 struct]
```

Then, referring to different parts of the variable would work the same:

```
>> myline.endpoint1
ans =
  x: 5
  y: 11
>> myline.endpoint2.x
ans =
  7
```

7.2.6 Vectors of Nested Structures

Combining vectors and nested structures, it is possible to have a vector of structures in which some fields are structures themselves. Here is an example in which a company manufactures cylinders from different materials for industrial use. Information on them is stored in a data structure in a program. The variable *cyls* is a vector of structures, each of which has fields *code, dimensions,* and *weight*. The *dimensions* field is a structure itself consisting of fields *rad* and *height* for the radius and height of each cylinder.

	code	dimensions		weight
		rad	height	
1	'x'	3	6	7
2	'a'	4	2	5
3	'c'	3	6	9

cyls

Here is an example of initializing the data structure by preallocating:

```
>> cyls(3) = struct('code', 'c', 'dimensions',...
   struct('rad', 3, 'height', 6), 'weight', 9);
>> cyls(1) = struct('code', 'x', 'dimensions',...
   struct('rad', 3, 'height', 6), 'weight', 7);
>> cyls(2) = struct('code', 'a', 'dimensions',...
   struct('rad', 4, 'height', 2), 'weight', 5);
```

Alternatively, it could be initialized by using the dot operator:

```
>> cyls(3).code = 'c';
>> cyls(3).dimensions.rad = 3;
>> cyls(3).dimensions.height = 6;
>> cyls(3).weight = 9;
>> cyls(1).code = 'x';
>> cyls(1).dimensions.rad = 3;
>> cyls(1).dimensions.height = 6;
>> cyls(1).weight = 7;
>> cyls(2).code = 'a';
>> cyls(2).dimensions.rad = 4;
>> cyls(2).dimensions.height = 2;
>> cyls(2).weight = 5;
```

There are several layers in this variable. For example,

- *cyls* is the entire data structure, which is a vector of **structs**

- *cyls(1)* is an individual element from the vector, which is a **struct**

- *cyls(2).code* is the *code* field from the **struct** *cyls(2)*; it is a character

- *cyls(3).dimensions* is the *dimensions* field from the **struct** *cyls(3)*; it is a **struct** itself

- *cyls(1).dimensions.rad* is the *rad* field from the **struct** *cyls(1).dimensions*; it is a **double** number

For these cylinders, one desired calculation may be the volume of each cylinder, which is defined as $\pi * r^2 * h$, where r is the radius and h is the height. The function *printcylvols* prints the volume of each cylinder, along with its *code* for identification purposes. It calls a subfunction to calculate each volume.

printcylvols.m

```
function printcylvols(cyls)
% This function prints the volumes of each cylinder
% It calls a subfunction to calculate each volume
for i = 1:length(cyls)
```

(Continued)

```
    vol = cylvol(cyls(i).dimensions);
    fprintf('Cylinder %c has a volume of %.1f\n', …
    cyls(i).code, vol);
  end

function cvol = cylvol(dims)
% Calculates the volume of a cylinder
cvol = pi * dims.rad ^ 2 * dims.height;
```

Here is an example of calling this function:

```
>> printcylvols(cyls)
Cylinder x has a volume of 169.6
Cylinder a has a volume of 100.5
Cylinder c has a volume of 169.6
```

Notice that the entire data structure, *cyls*, is passed to the function. The function loops through every element, each of which is a structure. It prints the code field for each, which is given by *cyls(i).code*. In order to calculate the volume of each cylinder, only the radius and height are needed, so rather than passing the entire structure to the subfunction *cylvol* (which would be *cyls(i)*), only the dimensions field is passed *(cyls(i). dimensions)*. The function then receives the *dimensions* structure as an input argument, and uses the dot operator to refer to the *rad* and *height* fields within it.

 PRACTICE 7.4

Modify the function *cylvol* to calculate the surface area of the cylinder in addition to the volume.

SUMMARY

Common Pitfalls

- Trying to use parentheses rather than curly braces for a cell array

- Forgetting to index into a vector using parentheses or referring to a field of a structure using the dot operator

Programming Style Guidelines

- Use arrays when values are the same type and represent in some sense the same thing.

- Use cell arrays or structures when the values are logically related but not the same type or the same thing.

- Use cell arrays rather than character matrices when storing strings of different lengths.

- Use cell arrays rather than structures when you want to loop through the values.

- Use structures rather than cell arrays when you want to use names for the different values rather than indices.

MATLAB Functions and Commands			
cell	cellstr	rmfield	fieldnames
celldisp	iscellstr	isstruct	
cellplot	struct	isfield	

MATLAB Operators	
cell arrays {}	dot operator for structs.

Exercises

1. Create a cell array that stores phrases, for example,

   ```
   exclaimcell = {'Bravo', 'Fantastic job'};
   ```

   Pick a random phrase to print.

2. Create the following cell array:

   ```
   >> ca = {'abc', 11, 3:2:9, zeros(2)}
   ```

   Use the **reshape** function to make it a 2 × 2 matrix. Then, write an expression that would refer to just the last column of this cell array.

3. Create a 2 × 2 cell array by using the **cell** function to preallocate and then put values in the individual elements. Then, insert a row in the middle so that the cell array is now 3 × 2. Hint: Extend the cell array by adding another row and copying row 2 to row 3, and then modify row 2.

4. Create three cell array variables that store people's names, verbs, and nouns. For example,

   ```
   names = {'Harry', 'Xavier', 'Sue'};
   verbs = {'loves', 'eats'};
   nouns = {'baseballs', 'rocks', 'sushi'};
   ```

   Write a script that will initialize these cell arrays, and then print sentences using one random element from each cell array, for example, 'Xavier eats sushi'.

5. Write a script that will prompt the user for strings and read them in, store them in a cell array (in a loop), and then print them out.

6. Create a row vector cell array to store the string 'xyz', the number 33.3, the vector 2:6, and the **logical** expression 'a' < 'c'. Use the transpose operator to make this a column vector, and use **reshape** to make it a 2 × 2 matrix. Use **celldisp** to display all elements.

7. Create a cell array variable that would store for a student his or her name, university ID number, and GPA. Print this information.

8. Create a structure variable that would store for a student his or her name, university ID number, and GPA. Print this information.

9. A complex number is a number of the form a + ib, where a is called the real part, b is called the imaginary part, and $i = \sqrt{-1}$. Write a script that prompts the user separately to enter values for the real and imaginary parts, and stores them in a structure variable. It then prints the complex number in the form a + ib. The script should just print the value of a, then the string '+ i', and then the value of b. For example, if the script is named *compnumstruct*, running it would result in:

```
>> compnumstruct
Enter the real part: 2.1
Enter the imaginary part: 3.3
The complex number is 2.1 + i3.3
```

(**Note**: This is just a structure exercise; MATLAB can handle complex numbers automatically as will be seen in Chapter 14.)

10. Modify the previous script to call a function to prompt the user for the real and imaginary parts of the complex number, and also call a function to print the complex number.

11. Write the code in MATLAB that would create the following data structure, and put the following values into the variable:

experiments						
num	**code**	**weights**		**height**		
		1	**2**	**feet**	**inches**	
1	33	'x'	200.34	202.45	5	6
2	11	't'	111.45	111.11	7	2

The variable is called *experiments*, which is a vector of structs. Each struct has four fields: *num, code, weights,* and *height*. The field *num* is an integer, *code* is a character, *weights* is a vector with two values (both of which are double values), and *height* is a struct with fields *feet* and *inches* (both of which are integers). Write the statements that would accomplish this, so that typing the following expressions in MATLAB would give the results shown:

```
>> experiments
experiments =
1x2 struct array with fields:
  num
  code
  weights
  height

>> experiments(2)
```

```
ans =
      num: 11
     code: 't'
  weights: [111.4500 111.1100]
   height: [1x1 struct]
>> experiments(1).height
ans =
     feet: 5
   inches: 6
```

12. Create a data structure to store information about the elements in the periodic table of elements. For every element, store the name, atomic number, chemical symbol, class, atomic weight, and a seven-element vector for the number of electrons in each shell. Create a structure variable to store the information, for example for lithium:

```
Lithium 3 Li alkali-metal 6.94 2 1 0 0 0 0 0
```

13. A team of engineers is designing a bridge to span the Podunk River. As part of the design process, the local flooding data must be analyzed. The following information on each storm that has been recorded in the last 40 years is stored in a file: the location of the source of the data, the amount of rainfall (in inches), and the duration of the storm (in hours), in that order. For example, the file might look like this:

```
321        2.4      1.5
111        3.3      12.1
           etc.
```

- Create a data file.
- Write the first part of the program: design a data structure to store the storm data from the file, and also the intensity of each storm. The intensity is the rainfall amount divided by the duration. Write a function to read the data from the file (use **load**), copy from the matrix into a vector of structs, and then calculate the intensities. Write another function to print all the information in a neatly organized table.
- Add a function to the program to calculate the average intensity of the storms.
- Add a function to the program to print all the information given on the most intense storm. Use a subfunction for this function that will return the index of the most intense storm.

14. A script stores information on potential subjects for an experiment in a vector of structures called *subjects*. The following show an example of what the contents might be:

```
>> subjects
subjects =
1x3 struct array with fields:
```

```
  name
  sub_id
  height
  weight
>> subjects(1)
ans =
name: 'Joey'
  sub_id: 111
  height: 6.7000
  weight: 222.2000
```

For this particular experiment, the only subjects who are eligible are those whose height or weight is lower than the average height or weight of all subjects. The script will print the names of those who are eligible. Create a vector with sample data in a script, and then write the code to accomplish this. Don't assume that the length of the vector is known; the code should be general.

15. Quiz data for a class is stored in a file. Each line in the file has the student ID number (which is an integer) followed by the quiz scores for that student. For example, if there are four students and three quizzes for each, the file might look like this:

44	7	7.5	8
33	5.5	6	6.5
37	8	8	8
24	6	7	8

First create the data file, and then store the data in a script in a vector of structures. Each element in the vector will be a structure that has two members: the integer student ID number and a vector of quiz scores. The structure will look like this:

| | students | | | |
| | id_no | quiz | | |
		1	2	3
1	44	7	7.5	8
2	33	5.5	6	6.5
3	37	8	8	8
4	24	6	7	8

To accomplish this, first use the **load** function to read all information from the file into a matrix. Then, using nested loops, copy the data into a vector of structures as specified previously. Then, the script will calculate and print the quiz average for each student; for example, if the file is as just shown:

Student	Quiz Ave
44	7.50
33	6.00
37	8.00
24	7.00

16. Create a nested struct to store a person's name, address, and phone numbers. The struct should have three fields for the name, address, and phone. The address fields and phone fields will be structs.

17. Design a nested structure to store information on constellations for a rocket design company. Each structure should store the constellation's name and information on the stars in the constellation. The structure for the star information should include the star's name, core temperature, distance from the sun, and whether it is a binary star or not. Create variables and sample data for your data structure.

18. Create a data file to store information on hurricanes. Each line in the file should have the name of the hurricane, its speed in miles per hour, and the diameter of its eye, in miles. Then, write a script to read this information from the file and create a vector of structures to store it. Print the name and area of the eye for each hurricane.

19. Create a data file to store blood donor information for a biomedical research company. For every donor, store their name, blood type, Rh factor, and blood pressure information. The blood type is either A, B, AB, or O. The Rh factor is + or –. The blood pressure consists of two readings: systolic and diastolic (both are **double** numbers). Write a script to read from your file into a data structure and print the information from the file.

20. Quality control involves keeping statistics on the quality of products. A company tracks its products and any failures that occur. For every imperfect part, a record is kept that includes the part number, a character code, a string that describes the failure, and the cost of both labor and material to fix the part. Create a vector of structures and create sample data for this company. Write a script that will print the information from the data structure in an easy-to-read format.

21. Create a data structure to store information on the planets in our solar system. For every planet, store its name, distance from the sun, and whether it is an inner planet or an outer planet.

22. A manufacturer is testing a new machine that mills parts. Several trial runs are made for each part, and the resulting parts that are created are weighed. A file stores, for every part, the part identification number, the ideal weight for the part, and also the weights from five trial runs of milling this part. Create a file in this format. Write a script that will read this information and store it in a vector of structures. For every part print whether the average of the trial weights was less than, greater than, or equal to the ideal weight.

23. Write a script that creates a vector of line segments (where each is a nested structure as shown in this chapter). Initialize the vector using any method. Print a table showing the values, for example,

```
Line     From        To
  1     ( 3, 5)    ( 4, 7)
  2     ( 5, 6)    ( 2, 10)
        etc.
```

24. Investigate the built-in function **cell2struct** that converts a cell array into a vector of structs.

25. Investigate the built-in function **struct2cell** that converts a struct to a cell array.

26. The following program stores information on boxes that have been made by the Acme Manufacturing Company. Unfortunately, many are rejects. The volume of each box is supposed to be above a certain threshold. The program checks this, and prints information on the boxes that do not meet this criterion. The program consists of a script that calls two functions: one that initializes a vector of nested structures with the box information, and a function that prints the rejects (this one calls a third function that calculates the volume of a box). Fill in the function that calculates the volume (length*width*height). For example, the output from the program might begin as:

```
>> acmebox
The following boxes are rejects:
Code Volume
   22  30.0
   33  6.0
```

acmebox.m

```
% This script initializes box structs and then calls a
% function to test for rejects based on a volume
% threshold of 55
box = initbox();
boxlims(box,55)
```

initbox.m

```
function box = initbox()
% Initializes a vector of box structs
box(1).code = 11;
box(1).dimensions.length = 3;
box(1).dimensions.width = 4;
box(1).dimensions.height = 5;
```

(Continued)

```
box(1).weight = 3.3;
box(2).code = 22;
% etc.
% Initialize several structures here but don't
% assume you know how many are initialized
```

boxlims.m

```
function boxlims(box, thresh)
% Prints the reject boxes
fprintf('The following boxes are rejects:\n')
fprintf('Code Volume\n')
for i = 1:length(box)
  bv = boxvol(box(i).dimensions);
  if bv < thresh
    fprintf(' %d %.1f\n', box(i).code,bv)
  end
end
```

Advanced File Input and Output

CONTENTS

This chapter extends the input and output concepts that were introduced in Chapter 2. In that chapter, we saw how to read values entered by the user using the **input** function, and also the output functions **disp** and **fprintf**, which display information in windows on the screen. For *file input and output* (file I/O), we used the **load** and **save** functions, which can read from a data file into a matrix, and write from a matrix to a data file. We also saw that there are three different operations that can be performed on files: reading from files, writing to files (implying writing to the beginning of a file), and appending to a file (writing to the end of a file).

There are many different *file types*, which use different filename extensions. So far, using **load** and **save**, we have worked with files in the ASCII format, which typically use either the extension .dat or .txt. The **load** command works only if there are the same number of values in each line and the values are the same type, so that the data can be stored in a matrix; the **save** command writes only from a matrix to a file. If the data to be written or file to be read is in a different format, *lower level file I/O functions* must be used.

The MATLAB® software has functions that can read and write data from different file types such as spreadsheets; typically Excel spreadsheets have the filename extension .xls. It also has its own binary file type, which uses the extension .mat. These are usually called MAT-files, and can be used to store variables that have been created in MATLAB.

In this chapter, we will introduce lower level file input and output functions, as well as functions that work with different file types.

8.1 LOWER LEVEL FILE I/O FUNCTIONS

When reading from a data file, the **load** function works as long as the data in the file is "regular"—in other words, the same kind of data on every line and in the same format on every line—so that it can be read into a matrix. However, data files are not always set up in this manner. When it is not possible to use **load**, MATLAB has what are called lower level file input functions that can be used. The file must be opened first, which involves finding or creating the file and positioning an indicator at the beginning of the file. When the reading has been completed, the file must be closed.

Similarly, the **save** function can write matrices to a file, but if the output is not a simple matrix there are lower level functions to write to files. Again, the file must be opened first and closed when the writing has been completed.

The steps involved are:

- Open the file.
- Read from the file, write to the file, or append to the file.
- Close the file.

First, the steps involved in opening and closing the file will be described. Several functions that perform the middle step of reading from or writing to the file will be described subsequently.

8.1.1 Opening and Closing a File

Files are opened with the **fopen** function. By default, the **fopen** function opens a file for reading. If another mode is desired, a permission string is used to specify which mode (e.g., writing or appending). The **fopen** function returns −1 if it is not successful in opening the file, or an integer value, which becomes the *file identifier* if it is successful. This file identifier is then used to refer to the file when calling other file I/O functions. The general form is:

```
fid = fopen('filename', 'permission string');
```

The permission strings include:

    r   reading (this is the default)
    w   writing
    a   appending

See **help fopen** for others.

After the **fopen** is attempted, the value returned should be tested to make sure that the file was successfully opened. For example, if the file does not exist, the **fopen** will not be successful. Since the **fopen** function returns −1 if the file was not found, this can be tested to decide whether to print an error message or to carry on and use the file. For example, if it is desired to read from a file 'samp. dat':

```
fid = fopen('samp.dat');
if fid == -1
    disp('File open not successful')
else
    % Carry on and use the file!
end
```

Files should be closed when the program has finished reading from or writing to them. The function that accomplishes this is the **fclose** function, which returns 0 if the file close was successful, or −1 if not. Individual files can be closed by specifying the file identifier, or if more than one file is open, all open files can be closed by passing the string 'all' to the **fclose** function. The general forms are:

```
closeresult = fclose(fid);
closeresult = fclose('all');
```

This should also be checked with an **if-else** statement to make sure it was successful.

8.1.2 Reading from Files

There are several lower level functions that read from files. The function **fscanf** reads formatted data into a matrix, using conversion formats such as %d for integers, %s for strings, and %f for floats (**double** values). The **textscan** function reads text data from a file and stores it in a cell array. The **fgetl** and **fgets** functions both read strings from a file one line at a time; the difference is that the **fgets** keeps the newline character if there is one at the end of the line, whereas the **fgetl** function gets rid of it. All these functions require first opening the file, and then closing it when finished.

Since the **fgetl** and **fgets** functions read one line at a time, these functions are typically in some form of a loop. The **fscanf** and **textscan** functions can read the entire data file into one data structure. In terms of level, these two functions are somewhat in between the **load** function and the lower level functions such as **fgetl**. The file must be opened using **fopen** first, and should be closed using **fclose** after the data has been read. However, no loop is required; they will read in the entire file automatically but into a data structure.

We will concentrate first on the **fgetl** function, which reads strings from a file one line at a time. The **fgetl** function affords more control over how the data is read than other input functions. The **fgetl** function reads one line of data from a file into a string; string functions can then be used to manipulate the data. Since **fgetl** reads only one line, it normally is placed in a loop that keeps going until the *end of the file* is reached. The function **feof** returns logical true if the end of the file has been reached. The function call **feof(fid)** would return logical true if the end of the file has been reached for the file identified by *fid*, or logical false if not. A general algorithm for reading from a file into strings would be:

- Attempt to open the file; check to make sure the file open was successful.

- If opened, loop until the end of the file is reached. For each line in the file,

  – read it into a string

  – manipulate the data

- Attempt to close the file; check to make sure the file close was successful.

The generic code to accomplish this is:

```
fid = fopen('filename');
if fid == -1
    disp('File open not successful')
else
    while feof(fid) == 0
        % Read one line into a string variable
        aline = fgetl(fid);
        % Use string functions to extract numbers, strings,
        % etc. from the line
        % Do something with the data!
    end
    closeresult = fclose(fid);
    if closeresult == 0
        disp('File close successful')
    else
        disp('File close not successful')
    end
end
```

The permission string could be included in the call to the **fopen** function, for example,

```
fid = fopen('filename', 'r');
```

but is not necessary since reading is the default. The condition on the **while** loop can be interpreted as saying "while the file end-of-file is false." Another way to write this is

```
while ~feof(fid)
```

which is interpreted as "while we're not at the end of the file."

For example, assume that there is a data file 'subjexp.dat', which has on each line a number followed by a character code. The **type** function can be used to display the contents of this file (since the file does not have the default extension .m, the extension on the filename must be included).

```
>> type subjexp.dat
5.3 a
2.2 b
3.3 a
4.4 a
1.1 b
```

The **load** function would not be able to read this into a matrix since it contains both numbers and text. Instead, the **fgetl** function can be used to read each line as a string and then string functions are used to separate the numbers and characters. For example, the following just reads each line and prints the number with 2 decimal places and then the rest of the string:

fileex.m

```
% Reads from a file one line at a time using fgetl
% Each line has a number and a character
% The script separates and prints them

% Open the file and check for success
fid = fopen('subjexp.dat');
if fid == -1
    disp('File open not successful')
else
    while feof(fid) == 0
        aline = fgetl(fid);
        % Separate each line into the number and character
```

(Continued)

```
            % code and convert to a number before printing
            [num charcode] = strtok(aline);
            fprintf('%.2f %s\n', str2num(num), charcode)
        end
        % Check the file close for success
        closeresult = fclose(fid);
        if closeresult == 0
            disp('File close successful')
        else
            disp('File close not successful')
        end
    end
```

Here is an example of executing this script:

```
>> fileex
5.30 a
2.20 b
3.30 a
4.40 a
1.10 b
File close successful
```

In this example, in the loop each time the **fgetl** function reads one line into a string variable. The string function **strtok** is then used to store the number and the character in separate variables, both of which are string variables. The function **str2num** must then be used to convert the number stored in the string variable into a **double** variable so that calculations could be performed on it.

 PRACTICE 8.1

Modify the script *fileex* to sum the numbers from the file. Create your own file in this format first.

Instead of using the **fgetl** function to read one line, once a file has been opened the **fscanf** function could be used to read in from this file directly into a matrix. However, the matrix must be manipulated somewhat to get it back into the original form from the file. The format of using the function is:

```
mat = fscanf(fid, 'format', [dimensions])
```

The **fscanf** reads into the matrix variable *mat* columnwise from the file identified by *fid*. The 'format' includes conversion characters much like those used in the **fprintf** function. The dimensions specify the desired dimensions of *mat*; if the number of values in the file is not known, **inf** can be used for the second dimension. For example, the following would read in the same file specified earlier, which has on each line a number, then a space, then a character.

```
>> fid = fopen('subjexp.dat');
>> mat = fscanf(fid,'%f %c',[2 inf])
mat =
      5.3000 2.2000 3.3000 4.4000 1.1000
      97.0000 98.0000 97.0000 97.0000 98.0000
>> fclose(fid);
```

The **fopen** opens the file for reading. The **fscanf** then reads from each line one double and one character, and places each pair in separate columns in the matrix. The dimensions specify that the matrix is to have two rows, by however many columns are necessary (equal to the number of lines in the file). Since matrices store values that are all the same type, the characters are stored as their ASCII equivalents in the character encoding (e.g., 'a' is 97). Once this matrix has been created, it may be more useful to separate the rows into vector variables and to convert the second back to characters, which can be accomplished as follows:

```
>> nums = mat(1,:);
>> charcodes = char(mat(2,:))
charcodes =
abaab
```

Of course, the results from **fopen** and **fclose** should be checked but were omitted here for simplicity.

 PRACTICE 8.2

Write a script to read in this file using **fscanf**, and sum the numbers.

QUICK QUESTION!

Instead of using the dimensions [2 inf] in the **fscanf** function, could we use
  [inf 2]?

Answer: No, [inf 2] would not work; since **fscanf** reads each row from the file into a column in the matrix. That means that the number of rows in the resulting matrix is known but the number of columns is not.

Another option for reading from a file is to use the **textscan** function. The **textscan** function reads text data from a file and stores it in a cell array. The **textscan** function is called, in its simplest form, as

```
cellarray = textscan(fid, 'format');
```

QUICK QUESTION!

Is the space in the conversion string '%f %c' important, or would the following also work?

```
>> mat = fscanf(fid,'%f %c',[2 inf])
```

Answer: No, that would not work. The conversion string '%f %c' specifies that there is a real number, then a space, then a character. Without the space in the conversion string, it would specify a real number immediately followed by a character (which would be the space in the file). Then, the next time it would be attempting to read the next real number but the file position indicator would be pointing to the character on the first line; the error would cause the **fscanf** function to halt. Here is the end result:

```
>> fid = fopen('subjexp.dat');
>> mat = fscanf(fid,'%f %c',[2 inf])
mat =
     5.3000
    32.0000
```

The 32 is the numerical equivalent of the space character ' ', as seen here.

```
>> double(' ')
ans =
    32
```

where the 'format' includes conversion characters much like those used in the **fprintf** function. For example, to read the file 'subjexp.dat' just shown, we could do the following (again, for simplicity, omitting the error-check of **fopen** and **fclose**):

```
>> fid = fopen('subjexp.dat');
>> subjdata = textscan(fid,'%f %c');
>> fclose(fid)
```

The format string '%f %c' specifies that on each line there is a **double** value followed by a space followed by a character. This creates a 1×2 cell array variable called *subjdata*. The first element in this cell array is a column vector of doubles (the first column from the file); the second element is a column vector of characters (the second column from the file), as shown here:

```
>> subjdata
subjdata =
    [5x1 double] [5x1 char]
>> subjdata{1}
ans =
     5.3000
     2.2000
     3.3000
     4.4000
     1.1000
>> subjdata{2}
ans =
a
b
```

```
a
a
b
```

To refer to individual values from the vector, it is necessary to index into the cell array using curly braces and then index into the vector using parentheses. For example, to refer to the third number in the first element of the cell array:

```
>> subjdata{1}(3)
ans =
    3.3000
```

A script that reads in this data and echo-prints it is shown here:

textscanex.m

```
% Reads data from a file using textscan
fid = fopen('subjexp.dat');
if fid == -1
    disp('File open not successful')
else
    % Reads numbers and characters into separate elements
    % in a cell array
    subjdata = textscan(fid,'%f %c');
    len = length(subjdata{1});
    for i= 1:len
        fprintf('%.1f %c\n',subjdata{1}(i),subjdata{2}(i))
    end
    closeresult = fclose(fid);
    if closeresult == 0
        disp('File close successful')
    else
        disp('File close not successful')
    end
end
```

Executing this script produces the following results:

```
>> textscanex
5.3 a
2.2 b
3.3 a
```

```
4.4 a
1.1 b
File close successful
```

PRACTICE 8.3

Modify the script *textscanex* to calculate the average of the column of numbers.

To summarize, we have now seen four methods of reading from a file. The function **load** will work only if the values in the file are all the same type and there is the same number on every line in the file, so that they can be read into a matrix. If this is not the case, lower level functions must be used. To use these, the file must be opened first and then closed when the reading has been completed. The **fscanf** function will read into a matrix, converting the characters to their ASCII equivalents. The **textscan** function will instead read into a cell array that stores each column from the file into separate column vectors of the cell array. Finally, the **fgetl** function can be used in a loop to read each line from the file as a separate string; string manipulating functions must then be used to break the string into pieces and convert to numbers.

QUICK QUESTION!

If a data file is in the following format, which file input function(s) could be used to read it in?

48	25	23	23
12	45	1	31
31	39	42	40

Answer: Any of the file input functions could be used, but since the file consists of only numbers, the **load** function would be the easiest.

8.1.3 Writing to Files

There are several lower level functions that can write to files. We will concentrate on the **fprintf** function, which can be used to write to a file and also to append to a file.

To write one line at a time to a file, the **fprintf** function can be used. Like the other low-level functions, the file must be opened first for writing (or appending) and should be closed once the writing has been completed. We have, of course, been using **fprintf** to write to the screen. The screen is the default output device, so if a file identifier is not specified, the output goes to the screen; otherwise, it goes to the specified file. The default file identifier number is 1 for the screen. The general form is:

```
fprintf(fid, 'format', variable(s));
```

The **fprintf** function actually returns the number of bytes that was written to the file, so if you do not want to see that number, suppress the output with a semicolon as shown here. (**Note:** When writing to the screen, the value returned by **fprintf** is not seen, but could be stored in a variable.)

Here is an example of writing to a file named 'tryit.txt':

```
>> fid = fopen('tryit.txt', 'w');
>> for i = 1:3
       fprintf(fid,'The loop variable is %d\n', i);
end
>> fclose(fid);
```

The permission string in the call to the **fopen** function specifies that the file is opened for writing to it. Just as when reading from a file, the results from **fopen** and **fclose** should really be checked to make sure they were successful. The **fopen** function attempts to open the file for writing. If the file already exists, the contents are erased so it is as if the file had not existed. If the file does not currently exist (which would be the norm), a new file is created. The **fopen** could fail, for example, if there isn't space to create this new file.

To see what was written to the file, we could then open it (for reading) and loop to read each line using **fgetl**:

```
>> fid = fopen('tryt.txt');
>> while ~feof(fid)
       aline = fgetl(fid)
    end
aline =
The loop variable is 1
aline =
The loop variable is 2
aline =
The loop variable is 3
>> fclose(fid);
```

Here is another example in which a matrix is written to a file. First, a random 2×4 matrix is created, and then it is written to a file using the format string '%d %d\n', which means that each column from the matrix will be written as a separate line in the file.

```
>> mat = randint(2,4,[5 20])
mat =
     20    14    19    12
      8    12    17     5
>> fid = fopen('randmat.dat','w');
>> fprintf(fid,'%d %d\n',mat);
>> fclose(fid);
```

Since this is a matrix, the **load** function can be used to read it in.

```
>> load randmat.dat
>> randmat
randmat =
     20     8
     14    12
     19    17
     12     5
>> randmat'
ans =
     20    14    19    12
      8    12    17     5
```

Transposing the matrix will display in the form of the original matrix. If you want this to begin with, the matrix variable *mat* can be transposed before using **fprintf** to write to the file. (Of course, it would be much simpler in this case to just use **save** instead!)

PRACTICE 8.4

Create a 3 × 5 matrix of random integers, each in the range from 1 to 100. Write this to a file called 'myrandmat.dat' in a 3 × 5 format using **fprintf**, so that the file appears identical to the original matrix. Load the file to confirm that it was created correctly.

8.1.4 Appending to Files

The **fprintf** function can also be used to append to an existing file. The permission string is 'a', for example,

```
fid = fopen('filename, 'a');
```

Unlike text files, data doesn't have to be in the same format as what is already in the file when appending.

8.2 WRITING AND READING SPREADSHEET FILES

The MATLAB functions **xlswrite** and **xlsread** will write to and read from spreadsheet files that have the extension .xls. For example, the following will create a 5 × 3 matrix of random integers, and then write it to a spreadsheet file called 'ranexcel.xls' that has five rows and three columns:

```
>> ranmat = randint(5,3,[1 100])
ranmat =
     96    77    62
     24    46    80
     61     2    93
     49    83    74
     90    45    18
>> xlswrite('ranexcel',ranmat)
```

The **xlsread** function will read from a spreadsheet file. For example, to read from the file just created:

```
>> ssnums = xlsread('ranexcel')
ssnums =
    96    77    62
    24    46    80
    61     2    93
    49    83    74
    90    45    18
```

In both cases the .xls extension on the filename is the default, so it can be omitted.

These are shown in their most basic forms, when the matrix or spreadsheet contains just numbers and the entire spreadsheet is read or matrix is written. There are many qualifiers that can be used for these functions, however. For example, the following would read from the spreadsheet file 'texttest.xls' that contains:

```
a         123       Cindy
b         333       Suzanne
c         432       David
d         987       Burt
```

```
>> [nums, txt] = xlsread('texttest.xls')
nums =
   123
   333
   432
   987
txt =
    'a'    ''    'Cindy'
    'b'    ''    'Suzanne'
    'c'    ''    'David'
    'd'    ''    'Burt'
```

This reads the numbers into a **double** vector variable *nums* and the text into a cell array *txt* (the **xlsread** function always returns the numbers first and then the text). The cell array is 4×3. It has three columns since the file had three columns, but since the middle column had numbers (which were extracted and stored in the vector *nums*), the middle column in the cell array *txt* consists of empty strings.

```
>> txt{1,2}
ans =
    ''
>> txt{1,3}
ans =
Cindy
```

A loop could then be used to echo-print the values from the spreadsheet in the original format:

```
>> for i = 1:length(nums)
      fprintf('%c %d %s\n', txt{i,1}, ...
         nums(i), txt{i,3})
   end
a 123 Cindy
b 333 Suzanne
c 432 David
d 987 Burt
```

8.3 USING MAT-FILES FOR VARIABLES

In addition to the data file types, MATLAB has functions that allow reading and saving variables from files. These files are called MAT-files (because the extension on the file name is .mat), and they store the names and contents of variables. Variables can be written to MAT-files, appended to them, and read from them.

Note: MAT-files are very different from the data files that we have worked with so far. Rather than just storing data, MAT-files store variable names and their values.

8.3.1 Writing Variables to a File

The **save** command can be used to write variables to a file, or to append variables to a MAT-file. By default, the **save** function writes to a MAT-file. It can either save the entire current workspace (all variables that have been created), or a subset of the workspace (including, for example, just one variable). The **save** function will save the file in the current directory, so it is important to set that correctly first.

To save all workspace variables in a file, the command is:

```
save filename
```

The .mat extension is added automatically to the filename. The contents of the file can be displayed using **who** with the -file qualifier:

```
who -file filename
```

For example, in the following session in the Command Window, three variables are created; these are then displayed using **who**. Then, the variables are saved to a file named 'sess1.mat'. The **who** function is then used to display the variables stored in that file.

```
>> mymat = rand(3,5)
mymat =
    0.9501    0.4860    0.4565    0.4447    0.9218
    0.2311    0.8913    0.0185    0.6154    0.7382
    0.6068    0.7621    0.8214    0.7919    0.1763
>> x = 1:6;
>> y = x.^2;
>> who
Your variables are:
mymat x y
>> save sess1
>> who -file sess1
Your variables are:
mymat x y
```

To save just one variable to a file, the format is:

```
save filename variablename
```

For example, just the matrix variable *mymat* is saved in a file called sess2:

```
>> save sess2 mymat
>> who -file sess2
Your variables are:
mymat
```

8.3.2 Appending Variables to a MAT-File

Appending to a file adds to what has already been saved in a file, and is accomplished using the –append option. For example, assuming that the variable *mymat* already has been stored in the file 'sess2.mat' as shown earlier, this would append the variable *x* to the file:

```
>> save -append sess2 x
>> who -file sess2
Your variables are:
mymat x
```

Without specifying variable(s), just **save –append** would add all variables from the Command Window to the file. When this happens, if the variable is not in the file, it is appended. If there is a variable with the same name in the file, it is replaced by the current value from the Command Window.

8.3.3 Reading from a MAT-File

The **load** function can be used to read from different types of files. As with the **save** function, by default the file will be assumed to be a MAT-file, and **load** can load all variables from the file or only a subset. For example, in a new

Command Window session in which no variables have been created yet, the **load** function could load from the files created in the previous section:

```
>> who
>> load sess2
>> who
Your variables are:
mymat x
```

A subset of the variables in a file can be loaded by specifying them in the form

```
load filename variable list
```

SUMMARY

Common Pitfalls

- Misspelling a filename, which causes a file open to be unsuccessful
- Using a lower level file I/O function, when **load** or **save** could be used
- Forgetting that **fscanf** reads columnwise into a matrix—so every line in the file is read into a column in the resulting matrix
- Forgetting that **fscanf** converts characters to their ASCII equivalents
- Forgetting that **textscan** reads into a cell array (so curly braces are necessary to index)
- Forgetting to use the permission string 'a' for appending to a file (which means the data already in the file would be lost!)

Programming Style Guidelines

- Use **load** when the file contains the same kind of data on every line and in the same format on every line.
- Always close files that were opened.
- Always check to make sure that files were opened and closed successfully.
- Make sure that all data is read from a file; for example, use a conditional loop to loop until the end of the file is reached rather than using a **for** loop.
- Be careful to use the correct formatting string when using **fscanf** or **textscan**.
- Store groups of related variables in separate MAT-files.

MATLAB Functions and Commands			
fopen	fgetl	textscan	xlsread
fclose	fgets	fprintf	
fscanf	feof	xlswrite	

Exercises

1. Write a script that will read from a file x and y data points in the following format:

```
x 0 y 1
x 1.3 y 2.2
x 2.2 y 6
x 3.4 y 7.4
```

The format of every line in the file is the letter x, a space, the x value, space, the letter y, space, and the y value. First, create the data file with 10 lines in this format. Do this by using the Editor/Debugger, then File Save As xypts.dat. The script will attempt to open the data file and error-check to make sure it was opened. If so, it uses a **for** loop and **fgetl** to read each line as a string. In the loop, it creates x and y vectors for the data points. After the loop, it plots these points and attempts to close the file. The script should print whether or not the file was successfully closed.

2. Modify the script from the previous problem. Assume that the data file is in exactly that format, but do not assume that the number of lines in the file is known. Instead of using a **for** loop, loop until the end of the file is reached. The number of points, however, should be in the plot title.

3. For a biomedical experiment, the names and weights of some patients have been stored in a file 'patwts.dat'. For example, the file might look like this:

```
Darby George 166.2
Helen Dee 143.5
Giovanni Lupa 192.4
Cat Donovan 215.1
```

Create this data file first. Then, write a script readpatwts that will first attempt to open the file. If the file open is not successful, an error message should be printed. If it is successful, the script will read the data into strings, one line at a time. Print for each person the name in the form 'last, first' followed by the weight. Also, calculate and print the average weight. Finally, print whether or not the file close was successful. For example, the result of running the script would look like this:

```
>> readpatwts
George,Darby 166.2
Dee,Helen 143.5
Lupa,Giovanni 192.4
Donovan,Cat 215.1
The ave weight is 179.30
File close successful
```

4. Create a file 'parts_inv.dat' that stores on each line a part number, cost, and quantity in inventory, for example:

```
123  5.99  52
456  3.97  100
333  2.22  567
```

Use **fscanf** to read this information, and print the total $ amount of inventory (the sum of the cost multiplied by the quantity for each part).

5. Create a file that stores on each line a letter, a space, and a real number. For example, it might look like this:

```
e  5.4
f  3.3
c  2.2
f  1.1
c  2.2
```

Write a script that uses **textscan** to read from this file. It will print the sum of the numbers in the file. The script should error-check the file open and close, and print error messages as necessary.

6. Create a file, 'phonenos.dat', of phone numbers in the following form:

```
6012425932
6178987654
8034562468
```

Read the phone numbers from the file and print them in the form:

```
601-242-5932
```

Use **load** to read the phone numbers.

7. Create the file 'phonenos.dat' as before. Use **textscan** to read the phone numbers, and then print them in the preceding format.

8. Create the file 'phonenos.dat' as before. Use **fgetl** to read the phone numbers in a loop, and then print them in the preceding format.

9. Write a script to read in division codes and sales for a company from a file that has the following format:

```
A.  4.2
B.  3.9
```

Print the division with the highest sales.

10. Modify any of the previous scripts to write the phone numbers in the new format to a new file.

11. Write a script that will loop to prompt the user for n circle radii. The script will call a function to calculate the area of each circle, and will write the results in sentence form to a file.

12. The Wind Chill Factor (WCF) measures how cold it feels with a given air temperature (T, in degrees Fahrenheit) and wind speed (V, in miles per hour). One formula for the WCF is

```
WCF = 35.7 + 0.6T - 35.7 (V^0.16) + 0.43T(V^0.16)
```

Create a table showing wind chill factors for temperatures ranging from –20 to 55 in steps of 5, and wind speeds ranging from 0 to 55 in steps of 5. Write this to a file 'wcftable.dat'.

13. Create a data file that has points in a three-dimensional space stored in the following format:

```
x 2.2 y 5.3 z 1.8
```

Do this by creating x, y, and z vectors and then use **fprintf** to create the file in the specified format.

14. Create a file that has some college department names and enrollments. For example, it might look like this:

```
Aerospace 201
Civil 45
Mechanical 66
```

Write a script that will read the information from this file and create a new file that has just the first four characters from the department names, followed by the enrollments. The new file will be in this form:

```
Aero 201
Civi 45
Mech 66
```

15. A software package writes data to a file in a format that includes curly braces around each line and commas separating the values. For example, a data file mm.dat might look like this:

```
{33, 2, 11}
{45, 9, 3}
```

Use the **fgetl** function in a loop to read this data. Create a matrix that stores just the numbers, and write the matrix to a new file. Assume that there are the same number of numbers in each line of the original file.

16. A file stores sales data (in millions) by quarters. For example, the format may look like this:

```
2007Q1 4.5
2007Q2 5.2
```

Create this file and then append the next quarter's data to it.

17. Create a spreadsheet that has on each line an integer student identification number followed by three quiz grades for that student. Read that information from the spreadsheet into a matrix, and print the average quiz score for each student.

18. The **xlswrite** function can write the contents of a cell array to a spreadsheet. A manufacturer stores information on the weights of some parts in a cell array. Each row stores the part identifier code followed by weights of some sample parts. To simulate this, create the following cell array:

```
>> parts = {'A22', 4.41 4.44 4.39 4.39
            'Z29', 8.88 8.95 8.84 8.92}
```

Then, write this to a spreadsheet file.

19. A spreadsheet, 'popdata.xls', stores the population every 20 years for a small town that underwent a boom and then a decline. Create this spreadsheet (include the header row) and then read the headers into a cell array and the numbers into a matrix. Plot the data using the header strings on the axis labels.

Year	Population
1920	4021
1940	8053
1960	14994
1980	9942
2000	3385

20. Create a multiplication table and write it to a spreadsheet.

21. Read numbers from any spreadsheet file, and write the variable to a MAT-file.

22. Clear out any variables that you have in your Command Window. Create a matrix variable and two vector variables.

   ■ Make sure that you have your current directory set.
   ■ Store all variables to a MAT-file.
   ■ Store just the two vector variables in a different MAT-file.
   ■ Verify the contents of your files using **who**.

23. Create a set of random matrix variables with descriptive names (e.g., *ran2by2int*, *ran3by3double*, etc.) for use when testing matrix functions. Store all of these in a MAT-file.

24. A data file is created as a **char** matrix and then saved to a file, for example,

```
>> cmat = char('hello', 'ciao', 'goodbye')
cmat =
hello
ciao
goodbye
>> save stringsfile.dat cmat -ascii
```

Can the **load** function be used to read this? What about **textscan**?

25. Create a file of strings as before, but create the file by opening a new M-file, type in strings, and then save it as a data file. Can the **load** function be used to read this? What about **textscan**?

Advanced Functions

CONTENTS

Functions were introduced in Chapter 2, and then expanded on in Chapter 5. In this chapter, several advanced features of functions and types of functions will be described. *Anonymous functions* are simple one-line functions that are called using their *function handle*. Other uses of function handles will also be demonstrated, including *function functions*. All the functions that we have seen so far have had a well-defined number of input and output arguments, but we will see that the number of arguments can be varied. *Nested functions* also are introduced, which are functions contained within other functions. Finally, *recursive functions* are functions that call themselves. A recursive function can return a value, or may simply accomplish a task such as printing.

9.1 ANONYMOUS FUNCTIONS

An anonymous function is a very simple, one-line function. The advantage of an anonymous function is that it does not have to be stored in an M-file. This can greatly simplify programs, since often calculations are very simple, and the use of anonymous functions reduces the number of M-files necessary for a program. Anonymous functions can be created in the Command Window or in any script. The syntax for an anonymous function is:

```
fnhandle = @ (arguments) functionbody
```

where *fnhandle* stores the **function handle**; it is essentially a way of referring to the function. The handle is assigned to this name using the @ operator. The arguments, in parentheses, correspond to the argument(s) that are passed to the function, just like any other kind of function. The functionbody is the body of the function, which is any valid MATLAB® expression. For example, here is an anonymous function that calculates and returns the area of a circle:

```
>> cirarea = @ (radius) pi * radius .^2;
```

The function handle name is *cirarea*. One argument is passed to the input argument *radius*. The body of the function is the expression pi * radius .^2. The .^ operator is used so that a vector of radii can be passed to the function.

The function is then called using the handle and passing arguments to it. The function call using the function handle looks just like a function call using a function name.

```
>> cirarea(4)
ans =
    50.2655
>> cirarea(1:4)
ans =
    3.1416  12.5664  28.2743  50.2655
```

Unlike functions stored in M-files, if no argument is passed to an anonymous function, the parentheses must still be in the function definition and in the function call. For example, here is an anonymous function that prints a random real number with two decimal places, as well as a call to this function:

```
>> prtran = @ () fprintf('%.2f\n',rand);
>> prtran()
0.95
```

Typing just the name of the function handle will display its contents, which is the function definition.

```
>> prtran
prtran =
        @ () fprintf('%.2f\n',rand)
```

This is why to call the function, even though no arguments are passed, the parentheses must be used.

To save an anonymous function, it can be saved to a MAT-file, and then it can be loaded when needed.

```
>> cirarea = @ (radius) pi * radius .^2;
>> save anonfns cirarea
>> clear
>> load anonfns
>> who
Your variables are:
cirarea

>> cirarea
cirarea =
        @ (radius) pi * radius .^2
```

Other anonymous functions could be appended to this MAT-file. Even though an advantage of anonymous functions is that they do not have to be saved in individual M-files, it is frequently useful to save groups of related anonymous functions in a MAT-file. If there are some anonymous functions that are frequently used, they can be saved in a MAT-file and then loaded from this MAT-file in every MATLAB Command Window.

9.2 USES OF FUNCTION HANDLES

Function handles can also be created for functions other than anonymous functions, both built-in and user-defined functions. For example, the following would create a function handle for the built-in **factorial** function:

```
>> facth = @factorial;
```

 PRACTICE 9.1

Create your own anonymous functions to calculate the areas of circles, rectangles, and something else (you decide!). Store these anonymous functions in a file called 'myareas.mat'.

The @ operator gets the handle of the function, which is then stored in a variable *facth*.

The handle could then be used to call the function, just like the handle for the anonymous functions, for example:

```
>> facth(5)
ans =
     120
```

Using the function handle to call the function instead of using the name of the function doesn't in itself demonstrate why this is useful, so an obvious question would be why function handles are necessary.

9.2.1 Function Functions

One reason for using function handles is to be able to pass functions to other functions—these are called *function functions*.

For example, let's say we have a function that creates an *x* vector. The *y* vector is created by evaluating a function at each of the x points, and then these points are plotted.

fnfnexamp.m

```
function fnfnexamp(funh)
% Example of a function function. The handle of a function
% is passed and that function of x is plotted
x = 1:.25:6;
y = funh(x);
plot(x,y,'ko')
```

FIGURE 9.1

Plot of **sin** *created by passing handle of function to plot.*

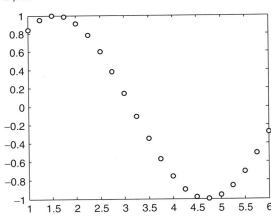

What we want is to be able to pass a function to be the value of *funh*, like **sin, cos, tan,** and so on. It doesn't work to just pass the name of the function:

```
>> fnfnexamp(sin)
??? Error using ==> sin
Not enough input arguments.
```

Instead, we have to pass the handle of the function:

```
>> fnfnexamp(@sin)
```

which creates the y vector as **sin(x)** and then brings up the plot as seen in Figure 9.1.

Passing instead the handle to the **cos** function would graph cosine instead of sine:

```
>> fnfnexamp(@cos)
```

We could also pass the handle of any user-defined or anonymous function to the *fnfnexamp* function.

QUICK QUESTION!

How could you write a function to accomplish essentially the same thing, but by passing the name of the function as a string?

Answer: Pass the name of the function as a string to the function, and then use **eval** to evaluate the function.

(Continued)

```
fnstrfn.m
```

```
function fnstrfn(funstr)
% A function name is passed as an argument to this
% function; it evaluates and plots the function of x
x = 1:.25:6;
y = eval(strcat(funstr,'(x)'));
plot(x,y,'bo')
```

In this case, the string '(x)' has to be concatenated to the name of the string to result in the string, for example, 'sin(x)', and then that is evaluated as a function using **eval**. This would be called by passing a string to the function, for example:

```
>> fnstrfn('sin')
```

This would create the same plot as shown in the previous example.

Another way of doing this is to use the built-in function **str2func** that will convert a string to a function handle:

```
fnstrfn2.m
```

```
function fnstrfn2(funstr)
% A function name is passed as an argument to this
% function; it converts this to a function handle and
% then plots the function of x
x = 1:.25:6;
funh = str2func(funstr);
y = funh(x);
plot(x,y,'bo')
```

This also would be called by passing a string to the function, and again would create the same plot:

```
>> fnstrfn2('sin')
```

There is also a function **func2str** that will convert a function handle to a string. This could be used to put the name of the function in the plot in Figure 9.1 (either in the title or on the y-axis label).

Although these methods also work, it seems simpler and cleaner to just pass the function handle to the function.

PRACTICE 9.2

Write a function that will receive as input arguments an *x* vector and a function handle, and will create a vector *y*, which is the function of *x* (whichever function handle is passed) and will also plot the data from the *x* and *y* vectors with the function name in the title.

MATLAB has some built-in function functions. One built-in function function is **fplot**, which plots a function between limits that are specified. The form of the call to **fplot** is:

```
fplot(fnhandle, [xmin xmax])
```

For example, to pass the **sin** function to **fplot** one would pass its handle (see Figure 9.2 for the result).

```
>> fplot(@sin, [-pi pi])
```

The **fplot** function is a nice shortcut—it is not necessary to create *x* and *y* vectors, and it plots a continuous curve rather than discrete points.

The function **feval** will evaluate a function handle and execute the function for the specified argument. For example, the following is equivalent to sin(3.2):

```
>> feval(@sin, 3.2)
ans =
      -0.0584
```

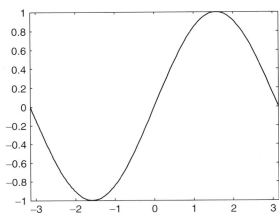

FIGURE 9.2

*Plot of **sin** created using* **fplot**.

9.3 VARIABLE NUMBERS OF ARGUMENTS

So far, in the functions that we've written, there has been a fixed number of input arguments and a fixed number of output arguments. For example, in the following function that we defined previously, there is one input argument and two output arguments:

```
areacirc.m
function [area, circum] = areacirc(rad)
% This function calculates the area and
% circumference of a circle
area = pi * rad .* rad;
circum = 2 * pi * rad;
```

However, this is not always the case. It is possible to have a *variable number of arguments*, both input arguments and output arguments. A built-in cell array **varargin** can be used to store a variable number of input arguments and a built-in cell array **varargout** can be used to store a variable number of output arguments. These are cell arrays because the arguments could be different types, and only cell arrays can store different kinds of values in the different elements. The function **nargin** returns the number of input arguments that were passed to the function, and the function **nargout** determines how many output arguments are expected to be returned from a function.

9.3.1 Variable Number of Input Arguments

For example, the following function *areafori* has a variable number of input arguments, either 1 or 2. The name of the function stands for "area, feet or inches." If only one argument is passed to the function, it represents the radius in feet. If two arguments are passed, the second can be a character 'i' indicating that the result should be in inches (for any other character, the default of feet is assumed). The function uses the built-in **varargin**, which stores as a cell array any number of input arguments. The function **nargin** returns the number of input arguments that were passed to the function. In this case, the radius is the first argument passed so it is stored in the first element in **varargin**. If a second argument is passed (if **nargin** is 2), it is to specify the units.

areafori.m

```
function area = areafori(varargin)
% Calculates and returns the area of a circle in feet
% The radius is passed, and potentially the unit of
% inches is also passed, in which case the result will be
% given in inches instead of feet
n = nargin; % number of input arguments
radius = varargin{1}; % Given in feet by default
if n == 2
    unit = varargin{2};
    % if inches is specified, convert the radius
    if unit == 'i'
        radius = radius * 12;
    end
end
area = pi * radius ^ 2;
```

Notice that the curly braces are used to refer to the elements in the cell array **varargin**.

Here are some examples of calling this function:

```
>> areafori(3)
ans =
```

```
    28.2743
>> areafori(1,'i')
ans =
   452.3893
```

In this case, it was assumed that the radius will always be passed to the function. The function header can therefore be modified to indicate that the radius will be passed, and then a variable number of remaining input arguments (either none or one):

areafori2.m

```
function area = areafori2(radius, varargin)
% Calculates and returns the area of a circle in feet
% The radius is passed, and potentially the unit of
% inches is also passed, in which case the result will be
% given in inches instead of feet
n = nargin; % number of input arguments
if n == 2
    unit = varargin{1};
    % if inches is specified, convert the radius
    if unit == 'i'
        radius = radius * 12;
    end
end
area = pi * radius ^ 2;
```

```
>> areafori2(1,'i')
ans =
     452.3893
>> areafori2(3)
ans =
     28.2743
```

Notice **nargin** returns the total number of input arguments, not just the number of arguments in the cell array **varargin**.

9.3.2 Variable Number of Output Arguments

A variable number of output arguments can also be specified. For example, one input argument is passed to the following function *typesize*. The function will always return a character specifying whether the input argument was a scalar

PRACTICE 9.3

The sum of a geometric series is given by

$$1 + r + r^2 + r^3 + r^4 + \ldots + r^n$$

Write a function called *geomser*, which will receive a value for *r*, and will calculate and return the sum of the geometric series. If a second argument is passed to the function, it is the value of *n*; otherwise, the function generates a random integer for *n* (in the range from 5 to 30). Note that loops are not necessary to accomplish this. The following examples of calls to this function illustrate what the result should be:

```
>> geomser(1,5) % 1 + 11 + 12 + 13 + 14 + 15
ans =
   6
>> g = geomser(2,4) % 1 + 21 + 22 + 23 + 24
g =
   31
>> geomser(1) % 1 + 11 + 12 + 13 +...?
ans =
   12
```

Note that in the last example, a random integer was generated for *n* (which must have been 11). Use the following header for the function, and fill in the rest:

```
function sgs = geomser(r, varargin)
```

('s'), vector ('v'), or matrix ('m'). This character is returned through the output argument *arrtype*. Additionally, if the input argument was a vector the function returns the length of the vector, and if the input argument was a matrix the function returns the number of rows and columns of the matrix. The output argument **varargout** is used; it is a cell array. So, for a vector the length is returned through **varargout** and for a matrix both the number of rows and columns are returned through **varargout**.

typesize.m

```
function [arrtype, varargout] = typesize(inputval)
% Demonstrates a variable number of output arguments
[r c ] = size(inputval);
if r==1 && c==1
    arrtype = 's';
```

(Continued)

```
elseif r==1 || c==1
     arrtype = 'v';
     varargout{1} = length(inputval);
else
     arrtype = 'm';
     varargout{1} = r;
     varargout{2} = c;
end
```

```
>> typesize(5)
ans =
s
>> [arrtype, len] = typesize(4:6)
arrtype =
v
len =
     3
>> [arrtype, r, c] = typesize([4:6;3:5])
arrtype =
m

r =
     2
c =
     3
```

In the examples shown here, the user must actually know the type of the argument in order to determine how many variables to have on the left-hand side of the assignment statement. An error will result if there are too many variables.

```
>> [arrtype, r,c] = typesize(4:6)
??? Error using ==> typesize
Too many output arguments.
```

The function **nargout** can be called to determine how many output arguments were used to call a function. For example, in the function *mysize*, next, a matrix is passed to the function. The function behaves like the built-in function **size** in that it returns the number of rows and columns. However, if three variables are used to store the result of calling this function, it also returns the total number of elements:

mysize.m

```
function [row col varargout] = mysize(mat)
% Demonstrates the use of nargout
[row col] = size(mat);
if nargout == 3
    varargout{1} = row*col;
end
```

```
>> [r c] = mysize(eye(3))
r =
    3
c =
    3
>> [r c elem] = mysize(eye(3))
r =
    3
c =
    3
elem =
    9
```

Notice that **nargout** does not return the number of output arguments in the function header, but the number of output arguments expected from the function (e.g., the number of arguments in the vector in the left-hand side of the assignment statement when calling the function). In the first call to the *mysize* function, the value of **nargout** was 2, so the function returned only the output arguments *row* and *col*. In the second call, since there were three variables on the left of the assignment statement, the value of **nargout** was 3 so the function also returned the total number of elements.

QUICK QUESTION!

A temperature in degrees C is passed to a function called *converttemp*. How could we write this function so that it converts this temperature to degrees F, and possibly also to degrees K, depending on the number of output arguments? The conversions are:

$$F = \frac{9}{5}C + 32$$

$$K = C + 273.15$$

Here are possible calls to the function:

```
>> df = converttemp(17)
df =
    62.6000
>> [df dk] = converttemp(17)
df =
    62.6000
dk =
    290.1500
```

(Continued)

Answer: We could write the function two different ways, using two different function headers: one that has just **varargout** in the function header, and one that has an output argument for the degrees F and also **varargout** in the function header.

converttemp.m

```
function [degreesF, varargout] = converttemp(degreesC)
% Convert temperature to degrees F and maybe also K
degreesF = 9/5*degreesC + 32;
n = nargout;
if n == 2
    varargout{1} = degreesC + 273.15;
end
```

converttempii.m

```
function varargout = converttempii(degreesC)
% Convert temperature to degrees F and maybe also K
n = nargout;
varargout{1} = 9/5*degreesC + 32;
if n == 2
    varargout{2} = degreesC + 273.15;
end
```

9.4 NESTED FUNCTIONS

Just as we have seen that loops can be nested, meaning one inside of another, functions can be nested. The terminology for *nested functions* is that an *outer function* can have within it *inner functions*. When functions are nested, every function must have an **end** statement (much like loops). The general format of a nested function is as follows:

```
outer function header
    body of outer function
    inner function header
        body of inner function
    end % inner function
    more body of outer function
end % outer function
```

The inner function can be in any part of the body of the outer function so there may be parts of the body of the outer function before and after the inner function. There can be multiple inner functions.

The scope of any variable is the workspace of the outermost function in which it is defined and used. That means that a variable defined in the outer function could be used in an inner function. A variable defined in the inner function *could* be used in the outer function, but if it is not used in the outer function the scope is just the inner function.

For example, the following function calculates and returns the volume of a cube. Three arguments are passed to it, for the length and width of the base of the cube and also the height. The outer function calls a nested function that calculates and returns the area of the base of the cube. Notice that it is not necessary to pass the length and width to the inner function, since the scope of these variables includes the inner function.

nestedvolume.m

```
function outvol = nestedvolume(len, wid, ht)
% Demonstrates a nested function
outvol = base * ht;
    function outbase = base
    outbase = len * wid;
    end % base function
end % nestedvolume function
```

Here is an example of calling this function:

```
>> v = nestedvolume(3,5,7)
v =
    105
```

Output arguments are different from variables. The scope of an output argument is just the nested function; it cannot be used in the outer function. In this example, *outbase* can be used only in the *base* function; its value, for example, could not be printed from *nestedvolume*.

Also, if a variable is not used in an outer function, then the variable is local to the nested function. The previous example has been modified to demonstrate these scope rules. In the following example, the value of the output argument *outbase* cannot be used by the function *nestedvolume2* (because it is an output argument, not a variable). Also, the variable *bvar* is local to the function *base* since it is never used in *nestedvolume2*. On the other hand, *cvar* is used by *nestedvolume2* (it is printed), so its scope is the workspace of *nestedvolume2*—meaning, for example, that it can be printed by *printstuff*.

nestedvolume2.m

```
function outvol = nestedvolume2(len, wid, ht)
% Demonstrates scope within a nested function
disp('This function calculates a volume')
% Call the base function, and calculate and
% print the volume
outvol = base * ht;
fprintf('outvol is %.1f\n', outvol)
fprintf('cvar is %.1f\n', cvar)
% Call the printstuff function
printstuff
% Not valid because it is an output argument:
% fprintf('outbase is %.1f\n', outbase)
    function outbase = base
        bvar = len * wid;
        cvar = len * wid;
        outbase = bvar;
    end
    function printstuff
        fprintf('outvol is %.1f\n', outvol)
        %Not valid because bvar is not used in nestedvolume2:
        % fprintf('bvar is %.1f\n', bvar)
        fprintf('cvar is %.1f\n', cvar)
    end
end
```

Here is an example of calling this modified function:

```
>> nestedvolume2(3,5,7)
This function calculates a volume
outvol is 105.0
cvar is 15.0
```

```
outvol is 105.0
cvar is 15.0
ans =
       105
```

9.5 RECURSIVE FUNCTIONS

Recursion occurs when something is defined in terms of itself. In programming, a *recursive function* is a function that calls itself. Recursion is used very commonly in programming, although many simple examples (including some shown in this section) are actually not very efficient and can be replaced by iterative methods (loops, or vectorized code in MATLAB). Nontrivial examples go beyond the scope of this book, so the concept of recursion is simply introduced here.

The example used will be of a factorial. Normally, the factorial of an integer n is defined iteratively:

```
n! = 1 * 2 * 3 *...* n
```

For example, `4! = 1 * 2 * 3 * 4`, or `24`.

Another, recursive, definition is:

```
n! = n * (n − 1)!        general case
1! = 1                   base case
```

This definition is recursive because a factorial is defined in terms of another factorial. There are two parts to any recursive definition: the general (or inductive) case, and the base case. We say that in general the factorial of n is defined as n multiplied by the factorial of (n – 1), but the base case is that the factorial of 1 is just 1. The base case stops the recursion.

For example,

```
3! = 3 * 2!
        2! = 2 * 1!
                1! = 1
           = 2
     = 6
```

The way this works is that 3! is defined in terms of another factorial, as 3 * 2!. This expression cannot yet be evaluated, because first we have to find out the value of 2!. So, in trying to evaluate the expression 3 * 2!, we are interrupted by the recursive definition. According to the definition, 2! is 2 * 1!. Again, the expression 2 * 1! cannot yet be evaluated because first we have to find the value of 1!. According to the definition, 1! is 1. Since we now know what 1! is, we can continue with the expression that was just being evaluated; now we know that 2 * 1! is 2 * 1, or 2. So, now we can finish the previous expression that was being evaluated; now we know that 3 * 2! is 3 * 2, or 6.

This is the way that recursion always works. With recursion, the expressions are put on hold with the interruption of the general case of the recursive definition. This keeps happening until finally the base case of the recursive definition applies. This finally stops the recursion, and then the expressions that were put on hold are evaluated in the reverse order. In this case, first the evaluation of 2 * 1! was completed, and then 3 * 2!.

There must always be a base case to end the recursion, and the base case must be reached at some point. Otherwise, *infinite recursion* would occur (theoretically, although MATLAB will stop the recursion eventually).

We have already seen the built-in function in MATLAB to calculate factorials, called **factorial** and we have seen how to implement the iterative definition using a running product. Now we will instead write a recursive function called **fact**. The function will receive an integer n, which for simplicity we will assume is a positive integer, and will calculate n! using the recursive definition just given:

fact.m

```
function facn = fact(n)
% This function recursively finds n!
if n == 1
    facn = 1;
else
    facn = n * fact(n-1);
end
```

The function calculates one value, using an **if-else** statement to choose between the base and general cases. If the value passed to the function is 1, the function returns 1 since 1! is equal to 1. Otherwise, the general case applies. According to the definition, the factorial of n, which is what this function is calculating, is defined as n multiplied by the factorial of (n–1). So, the function assigns n * fact(n–1) to the output argument.

How does this work? Exactly the way the example was sketched for 3!. Let's trace what would happen if the integer 3 is passed to the function:

```
fact(3) tries to assign 3 * fact(2)
        fact(2) tries to assign 2 * fact(1)
                fact(1) assigns 1
        fact(2) assigns 2
fact(3) assigns 6
```

When the function is first called, 3 is not equal to 1, so the statement

```
facn = n * fact(n-1);
```

is executed. This will attempt to assign the value of 3 * fact(2) to *facn*, but this expression cannot be evaluated yet and therefore a value cannot be assigned yet because first the value of fact(2) must be found. Thus, the assignment statement has been interrupted by a recursive call to the *fact* function. The call to the function *fact(2)* results in an attempt to assign 2 * fact(1), but again this expression cannot yet be evaluated. Next, the call to the function *fact(1)* results in a complete execution of an assignment statement since it assigns just 1. Once the base case has been reached, the assignment statements that were interrupted can be evaluated, in the reverse order.

Calling this function yields the same result as the built-in **factorial** function:

```
>> fact(5)
ans =
     120
>> factorial(5)
ans =
     120
```

The recursive factorial function is a very common example of a recursive function. It is somewhat of a lame example, however, since recursion is not necessary to find a factorial; a **for** loop can be used just as well in programming (or, of course, the built-in function in MATLAB).

Another, better, example is of a recursive function that does not return anything, but simply prints. The following function *prtwords* receives a sentence, and prints the words in the sentence in reverse order. The algorithm for the *prtwords* function is:

- Receive a sentence as an input argument.
- Use **strtok** to break the sentence into the first word and the rest of the sentence.
- If the rest of the sentence is not empty (in other words, if there is more to it), recursively call the *prtwords* function and pass to it the rest of the sentence.
- Print the word.

The function definition is:

prtwords.m

```
function prtwords(sent)
% This function recusively prints the words in a string
% in reverse order
[word, rest] = strtok(sent);
if ~isempty(rest)
    prtwords(rest);
end
disp(word)
```

Here is an example of calling the function, passing the sentence "what does this do":

```
>> prtwords('what does this do')
do
this
does
what
```

What happens when the function is called here is outlined as follows.

```
The function receives 'what does this do'
It breaks it into word = 'what', rest = ' does this do'
Since rest is not empty, calls prtwords, passing rest
    The function receives ' does this do'
    It breaks it into word = 'does', rest = ' this do'
    Since rest is not empty, calls prtwords, passing rest
        The function receives ' this do'
        It breaks it into word = 'this', rest = ' do'
        Since rest is not empty, calls prtwords, passing rest
            The function receives ' do'
            It breaks it into word = 'do', rest = ''
            rest is empty so no recursive call
            Print 'do'
        Print 'this'
    Print 'does'
Print 'what'
```

In this example, the base case is when the rest of the string is empty; in other words, the end of the original sentence has been reached. Every time the function is called, the execution of the function is interrupted by a recursive call to the function, until the base case is reached. When the base case is reached, the entire function can be executed, including the printing of the word (in the base case, the word 'do'). Once that execution of the function is completed, the program returns to the previous version of the function in which the word was 'this', and finishes the execution by printing the word 'this'. This continues; the versions of the function are finished in the reverse order, so the program ends up printing the words from the sentence in the reverse order.

 PRACTICE 9.4

For the following function:

recurfn.m

```
function outvar = recurfn(num)
if num < 0
   outvar = 4;
else
   outvar = 3 + recurfn(num-1);
end
```

What would be returned by the call to the function *recurfn(2.3)*? Think about it, and then type the function and test it.

SUMMARY

Common Pitfalls

- Trying to pass just the name of a function to a function function; instead, the function handle must be passed

- Thinking that **nargin** is the number of elements in **varargin** (it is not necessarily; it is the total number of input arguments)

- Forgetting the base case for a recursive function

Programming Style Guidelines

- Use anonymous functions whenever the function body consists of just a simple expression.

- Store related anonymous functions together in one MAT-file.

- If some inputs and outputs will always be passed to/from a function, use standard input arguments/output arguments for them. Use **varargin** and **varargout** only when it is not known ahead of time whether other input/output arguments will be needed.

- Use iteration instead of recursion when possible.

MATLAB Functions and Commands			
str2func	feval	nargin	end (for
func2str	varargin	nargout	functions)
fplot	varargout		

Exercises

1. An approximation for a factorial can be found using Stirling's formula:

$$n! \approx \sqrt{2\pi n}\left(\frac{n}{e}\right)^{n}$$

   Write an anonymous function to implement this.

2. Write an anonymous function to calculate and return the area of a rectangle.

3. The velocity of sound in air is $49.02\sqrt{T}$ feet per second where T is the air temperature in degrees Rankine. Write an anonymous function that will calculate this. One argument, the air temperature in degrees R, will be passed to the function and it will return the velocity of sound.

4. The hyperbolic sine for an argument x is defined as:

   ```
   hyperbolicsine(x) = (eˣ − e−ˣ)/ 2
   ```

   Write an anonymous function to implement this. Compare yours to the built-in function **sinh**.

5. Create a set of anonymous functions to do length conversions and store them in a file lenconv.mat. Call them a descriptive name, for example, *cmtoinch* to convert from centimeters to inches.

6. Write a function that will receive data in the form of x and y vectors, and a handle to a plot function and will produce the plot. For example, a call to the function would look like wsfn(x,y,@bar).

7. Write a function *plot2fnhand* that will receive two function handles as input arguments, and will display in two Figure Windows plots of these functions, with the function names in the titles. The function will create an x vector that ranges from 1 to n (where n is a random integer in the range from 4 to 10). For example, if the function is called as follows:

   ```
   >> plot2fnhand(@sqrt, @exp)
   ```

   and the random integer is 5, the Figure Window 1 would display the **sqrt** function of x from 1 to 5, and the second Figure Window would display **exp(x)** for x = 1:5.

8. Write an anonymous function to implement the following quadratic: $3x^2 - 2x + 5$. Then, use **fplot** to plot the function in the range from –6 to 6.

9. Use **feval** as an alternative way to accomplish the following function calls:

   ```
   abs(-4)

   size(zeros(4)) (Use feval twice for this one!)
   ```

10. There is a built-in function function called **cellfun** that evaluates a function for every element of a cell array. Create a cell array, then call the **cellfun** function, passing the handle of the **length** function and the cell array in order to determine the length of every element in the cell array.

11. Write a function that will print a random integer. If no arguments are passed to the function, it will print an integer in the range from 1 to 100. If one argument is passed, it is the max and the integer will be in the range from 1 to max. If two arguments are passed, they represent the min and max and it will print an integer in the range from min to max.

12. The velocity of sound in air is $49.02 \sqrt{T}$ feet per second where T is the air temperature in degrees Rankine. Write a function to implement this. If just one argument is passed to the function, it is assumed to be the air temperature in degrees Rankine. If, however, two arguments are passed, the two arguments would be first an air temperature and then a character 'f' for Fahrenheit or 'c' for Celsius (so this would then have to be converted to Rankine). **Note:** degrees R = degrees F + 459.67. Degrees F = 9/5 degrees C + 32.

13. Write a function *areaperim* that will calculate both the area and perimeter of a polygon. The radius r will be passed as an argument to the function. If a second argument is passed to the function, it represents the number of sides n. If, however, only one argument is passed, the function generates a random value for n (an integer in the range from 3 to 8). For a polygon with n sides inscribed in a circle with a radius of r, the area a and perimeter p of the polygon can be found by

$$a = \frac{1}{2} nr^2 \sin\left(\frac{\pi}{n}\right)$$

$$p = 2\pi\, r \sin\left(\frac{\pi}{n}\right)$$

14. Write a function that will receive a variable number of input arguments: the length and width of a rectangle, and possibly also the height of a box that has this rectangle as its base. The function should return the rectangle area if just the length and width are passed, or also the volume if the height is also passed.

15. Write a function that will receive the radius r of a sphere. It will calculate and return the volume of the sphere (4/3 π r³). If the function call expects two output arguments, the function will also return the surface area of the sphere (4 π r²).

16. The built-in function **clock** returns a vector with six elements representing, in order, the year, month, day, hour, minutes, and seconds. Write a function *whatday* that (using the **clock** function) will always return the current day. If the function call expects two output arguments, it will also return the month. If the function call expects three output arguments, it will also return the year.

17. The built-in function **date** returns a string containing the day, month, and year. Write a function (using the **date** function) that will always return the current day. If the function call expects two output arguments, it will also return the month. If the function call expects three output arguments, it will also return the year.

18. Write a function to calculate the volume of a cone. The volume V is V = AH, where A is the area of the circular base (A = πr² where r is the radius) and H is the height. Use a nested function to calculate A.

19. The two real roots of a quadratic equation $ax^2 + bx + c = 0$ (where a is nonzero) are given by

$$\frac{-b \pm \sqrt{D}}{2 * a}$$

where the discriminant $D = b^2 - 4 * a * c$. Write a function to calculate and return the roots of a quadratic equation. Pass the values of a, b, and c to the function. Use a nested function to calculate the discriminant.

20. A recursive definition of a^n where **a** is an integer and **n** is a nonnegative integer is:

```
aⁿ =  1     if n == 0
   = a * aⁿ⁻¹    if n > 0
```

Write a recursive function called *mypower*, which receives **a** and **n** and returns the value of a^n by implementing the above definition. Note: The program should **not** use ^ operator anywhere; this is to be done recursively instead! Test the function.

21. What does this function do?

```
function outvar = mystery(x,y)
if y == 1
  outvar = x;
else
  outvar = x + mystery(x,y-1);
end
```

Give one word to describe what this function does with its two arguments.

22. The Fibonacci numbers is a sequence of numbers F_i:

```
0 1 1 2 3 5 8 13 21 34 ...
```

where F_0 is 0, F_1 is 1, F_2 is 1, F_3 is 2, and so on. The sequence starts with 0 and 1. All other Fibonacci numbers are obtained by adding the previous two Fibonacci numbers. A recursive definition is:

```
F₀ = 0
F₁ = 1
Fₙ = Fₙ₋₂ + Fₙ₋₁ if n > 1
```

Write a recursive function to implement this definition. The function will receive one integer argument n, and it will return one integer value, which is the nth Fibonacci number. Note that in this definition there is one general case but two base cases. Then, test the function by printing the first 20 Fibonacci numbers.

23. Use **fgets** to read strings from a file and recursively print them backward.

24. Combinatorial coefficients can be defined recursively as follows:

```
C(n,m) = 1 if m = 0 or m = n
       = C(n-1, m-1) + C(n-1, m)otherwise
```

Write a recursive function to implement this definition.

PART

2

Applications

CHAPTER 10

MATLAB Plots

KEY WORDS

histogram	bin	object handle
stem plot	animation	hyteograph
pie chart	plot property	inverse functions
area plot	object	

CONTENTS

In Chapter 2, we introduced the use of the function **plot** in the MATLAB® software to get simple, two-dimensional plots of x and y points represented by two vectors, *x* and *y*. We also have seen some functions that allow customization of these plots. In this chapter we will explore other types of plots, ways of customizing plots, and some applications that combine plotting with functions and file input. Additionally, animation, three-dimensional plots, and graphics properties will be introduced.

10.1 PLOT FUNCTIONS

So far, we have used **plot** to create two-dimensional plots and **bar** to create bar charts. We have seen how to clear the Figure Window using **clf**, and how to create and number Figure Windows using **figure**. Labeling plots has been accomplished using **xlabel**, **ylabel**, **title**, and **legend**, and we also have seen how to customize the strings passed to these functions using **sprintf**. The **axis** function changes the axes from the defaults that would be taken from the data in the *x* and *y* vectors to the values specified. Finally, the **grid** and **hold** toggle functions print grids or not, or lock the current graph in the Figure Window so that the next plot will be superimposed.

10.1.1 Matrix of Plots

Another function that is very useful with any type of plot is **subplot**, which creates a matrix of plots in the current Figure Window. Three arguments are passed to it in the form **subplot(r,c,n)**; where *r* and *c* are the dimensions of the matrix and *n* is the number of the particular plot within this matrix. The plots are numbered rowwise starting in the upper left corner. In many cases, it is useful to create a **subplot** in a **for** loop so the loop variable can iterate through the integers 1 through *n*. When the **subplot** function is called in a loop, the first two arguments will always be the same since they give the dimensions of the matrix. The third argument will iterate through the numbers assigned to the elements of the matrix. When the **subplot** function is called, it makes that element the active plot; then, any plot function can be used complete with axis labeling, titles, and such within that element.

FIGURE 10.1

Subplot to demonstrate a plot using 10 points and 20 points.

For example, the following **subplot** shows the difference, in one Figure Window, between using 10 points and 20 points to plot **sin(x)** between 0 and 2 * π. The **subplot** function creates a 1 × 2 row vector of plots in the Figure Window, so that the two plots are shown side-by-side. The loop variable *i* iterates through the values 1 and then 2. The first time through the loop, when *i* has the value 1, 10*1 or 10 points are used, and the value of the third argument to the **subplot** function is 1. The second time through the loop, 20 points are used and the third argument to **subplot** is 2. Note that **sprintf** is used to print how many points were used in the plot titles. The resulting Figure Window with both plots is shown in Figure 10.1.

```
subplotex.m
```

```
%demonstrates subplot using a for loop
for i = 1:2
  x = linspace(0,2*pi,10*i);
  y = sin(x);
```

(Continued)

```
    subplot(1,2,i)
    plot(x,y,'ko')
    ylabel('sin(x)')
    title(sprintf('%d Points',10*i))
end
```

10.1.2 Plot Types

Besides **plot** and **bar**, there are other plot types such as *histograms, stem plots, area plots* and *pie charts*, as well as other functions that customize graphs.

Described in this section are some of the other plotting functions. The functions **bar**, **barh**, **area**, and **stem** essentially display the same data as the **plot** function, but in different forms. The **bar** function draws a bar chart (as we have seen before), **barh** draws a horizontal bar chart, **area** draws the plot as a continuous curve and fills in under the curve that is created, and **stem** draws a stem plot.

For example, the following script creates a Figure Window that uses a 2 × 2 **subplot** to demonstrate these four plot types using the same x and y points (see Figure 10.2).

FIGURE 10.2
Subplot to display **bar**, **barh**, *area, and* **stem** *plots.*

subplottypes.m

```
% Subplot to show plot types
x = 1:6;
y = [33 11 5 9 22 30];
subplot(2,2,1)
bar(x,y)
title('bar')
subplot(2,2,2)
barh(x,y)
title('barh')
```

(Continued)

```
subplot(2,2,3)
area(x,y)
title('area')
subplot(2,2,4)
stem(x,y)
title('stem')
```

Notice that the third argument in the call to the **subplot** function is a single index into the matrix created in the Figure Window; the numbering is rowwise (in contrast to the normal columnwise unwinding that MATLAB uses for matrices).

QUICK QUESTION!

Could we produce this **subplot** using a loop?
Answer: Yes, we can store the names of the plots in a cell array. These names are put in the titles, and also concatenated with the string '(x,y)' and passed to the **eval** function to evaluate the function.

loopsubplot.m

```
% Demonstrates evaluating plot type names in order to
% use the plot functions and put the names in titles
x = 1:6;
y = [33 11 5 9 22 30];
titles = {'bar', 'barh', 'area', 'stem'};
for i = 1:4
  subplot(2,2,i)
  eval([titles{i} '(x,y)'])
  title(titles{i})
end
```

For a matrix, the **bar** and **barh** functions will group together the values in each row. For example:

```
rm = randint(2,4,[1 50])
rm =
            8          19         43          25
           35          44         30          45
bar(rm)
```

produces the plot shown in Figure 10.3.

Notice that MATLAB groups together the values in the first row and then in the second row. It cycles through colors to distinguish the bars. The 'stack' option will stack rather than group the values, so the y value represented by the top of the bar is the sum of the values from that row (seen in Figure 10.4).

```
bar(rm,'stack')
```

A *histogram* is a particular type of bar chart that shows the frequency of occurrence of values within a vector. Histograms use what are called *bins* to collect values that are in given ranges. MATLAB has a function to create a histogram, **hist**. Calling the function with the form **hist(vec)** by default takes the values in the vector *vec* and puts them into 10 bins (or, **hist(vec,n)** will put them into n bins) and plots this, as seen in Figure 10.5.

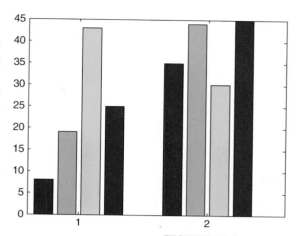

FIGURE 10.3
Data from a matrix in a bar chart.

```
>> vec = randint(1,12,[1 10])
vec =
10  8  2  5  10  10  5  9  1  4  9  1
>> hist(vec)
```

In this example, the numbers range from 1 to 10 in the vector, so there are 10 bins in the range from 1 to 10. The heights of the bins represent the number of values that fall within that particular bin. The **hist** function actually returns values; the first returned is a vector showing how many of the values from the original vector fall into each of the bins:

```
>> c = hist(vec)
c =
2  1  0  1  2  0  0  1  2  3
```

FIGURE 10.4
Stacked bar chart of matrix data.

The bins in a histogram are not necessarily all the same width. Histograms are used for statistical analyses on data; more statistics will be covered in Chapter 12.

MATLAB also has a function **pie** that will create a pie chart. Calling the function with the form **pie(vec)** draws a pie chart, using the percentage of each element of *vec* of the whole (the sum). It shows these starting from the top of the circle and going around counterclockwise. For example, the first value in the vector

---

 PRACTICE 10.1

Create a file that has two lines with *n* numbers in each. Use **load** to read into a matrix. Then, use **subplot** to show the **bar** and stacked **bar** charts side-by-side.

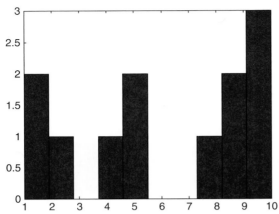

FIGURE 10.5
Histogram of data.

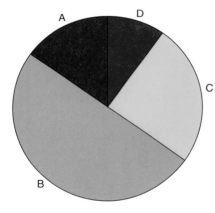

FIGURE 10.6
Pie chart showing percentages.

FIGURE 10.7
Pie chart with labels from a cell array.

[3 10 5 2] , 3, is 15% of the sum, 10 is 50% of the sum, and so forth, as seen in Figure 10.6.

```
>> pie ([3 10 5 2])
```

A cell array of labels can also be passed to the **pie** function; these labels will appear instead of the percentages (seen in Figure 10.7).

```
>> pie([3 10 5 2], {'A','B','C','D'})
```

10.2 ANIMATION

In this section we will examine a couple of ways to *animate* a plot. These are visuals, so the results can't really be shown here; it is necessary to type these into MATLAB to see the results.

We'll start by *animating* a plot of **sin(x)** with the vectors:

```
>> x = -2*pi : 1/100 : 2*pi;
>> y = sin(x);
```

This results in enough points that we'll be able to see the result using the built-in **comet** function,

 PRACTICE 10.2

A chemistry professor teaches three classes. These are the course numbers and enrollments:

```
CH 101              111
CH 105               52
CH 555               12
```

Use **subplot** to show this information using **pie** charts: the **pie** chart on the left should show the percentage of students in each course, and on the right, the course numbers. Put appropriate titles on them.

which shows the plot by first showing the point (x(1),y(1)), and then moving on to the point (x(2),y(2)), and so on, leaving a trail (like a comet!) of all the previous points.

```
>> comet(x,y)
```

The end result looks the same as **plot(x,y)**.

Another way of animating is to use the built-in function **movie**, which displays recorded movie frames. The frames are captured in a loop using the built-in function **getframe**, and are stored in a matrix. For example, the following script again animates the **sin** function. The **axis** function is used so that MATLAB will use the same set of axes for all frames, and using the **min** and **max** functions on the data vectors x and y will allow us to see all points. It displays the movie once in the **for** loop, and then again when the **movie** function is called.

sinmovie.m

```
% Shows a movie of the sin function
clear
x = -2*pi: 1/5 : 2*pi;
y = sin(x);
n = length(x);
for i = 1:n
    plot(x(i),y(i),'r*')
    axis([min(x)-1 max(x)+1 min(y)-1 max(y)+1])
    M(i) = getframe;
end
movie(M)
```

10.3 THREE-DIMENSIONAL PLOTS

MATLAB has many functions that will display three-dimensional plots. Most of these functions have the same name as the corresponding two-dimensional plot function with a 3 at the end. For example, the three-dimensional line plot function is called **plot3**. Other functions include **bar3**, **pie3**, and **stem3**.

Vectors representing x, y, and z coordinates are passed to the **plot3** and **stem3** functions. These functions show the points in three-dimensional space. Clicking on the rotate 3D icon and then in the plot allows the user to rotate the view to see the plot from different angles. Also, using the **grid** function makes it easier to visualize, as seen in Figure 10.8.

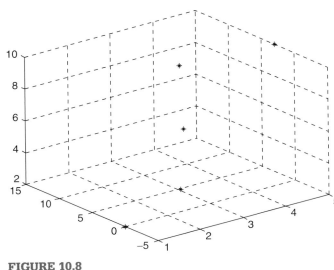

```
>> x = 1:5;
>> y = [0 -2 4 11 3];
>> z = 2:2:10;
>> plot3(x,y,z,'k*')
>> grid
```

For the **bar3** function, x and y vectors are passed and the function shows three-dimensional bars as seen in Figure 10.9.

```
>> x = 1:6;
>> y = [33 11 5 9 22 30];
>> bar3(x,y)
```

Similarly, the **pie3** function shows data from a vector as a three-dimensional pie as seen in Figure 10.10.

```
>> pie3([3 10 5 2])
```

FIGURE 10.8
Three-dimensional plot with a grid.

10.4 CUSTOMIZING PLOTS

There are many ways to customize figures in the Figure Window. Clicking on the Plot Tools icon will bring up the Property Editor and Plot Browser, with many options for modifying the current plot. Additionally, there are **properties** that can be modified from the default in the **plot** function itself. Using the **help** facility with the function name will show all the options for that particular plot function.

For example, the **bar** and **barh** functions by default put a width of 0.8 between bars. When called as **bar(x,y)**, the width of 0.8 is used. If instead a third argument is passed, it is the width, for example, **bar(x,y, width)**. The following script uses **subplot** to show variations on the width. A width of 0.6 results in more space between the bars. A width of 1 makes the bars touch each other, and with a width greater than 1.2, the bars actually overlap. The results are seen in Figure 10.11.

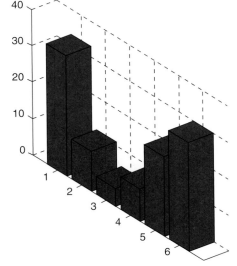

FIGURE 10.9
Three-dimensional bar chart.

FIGURE 10.10
Three-dimensional pie chart.

FIGURE 10.11

Subplot demonstrates varying widths in a bar chart.

barwidths.m

```
% Subplot to show varying bar widths
x = 1:6;
y = [33 11 5 9 22 30];
for i = 1:4
    subplot(1,4,i)
    % width will be 0.6, 0.8, 1, 1.2
    barh(x,y,0.4+i*.2)
    title(sprintf('Width = %.1f',0.4+i*.2))
end
```

PRACTICE 10.3

Use **help area** to find out how to change the base level on an **area** chart.

As another example of customizing plots, pieces of a pie chart can be "exploded" from the rest. In this case, two vectors are passed to the **pie** function: first the data vector, then a logical vector; the elements for which the logical vector is true will be exploded from (separated from) the pie chart. A third argument, a cell array of labels, can also be passed. The result is seen in Figure 10.12.

FIGURE 10.12

Exploding pie chart.

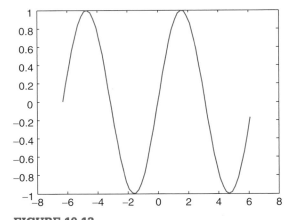

FIGURE 10.13

Plot of **sin** *function with default properties.*

```
>> vec = [3 10 5 2];
>> which = (vec == 5)
which =
              0    0    1    0

>> pie(vec, which, {'A', 'B', 'C', 'D'})
```

10.5 GRAPHICS PROPERTIES

MATLAB uses what it calls Handle Graphics® in all its figures. All figures consist of different **objects**, each of which is assigned a **handle**. The object handle is a unique real number that is used to refer to the object.

The various plot functions return a handle for the plot object, which can then be stored in a variable. In the following, the **plot** function plots a **sin** function in a Figure Window (as seen in Figure 10.13) and returns a real number, which is the object handle. (Don't try to make sense of the actual number used for the handle!) This handle will remain valid as long as the object exists.

```
>> x = -2*pi: 1/5 : 2*pi;
>> y = sin(x);
>> h1 = plot(x,y)
h1 =
      158.5420
```

Note: After getting the plot, the Figure Window should not be closed, as that would make the object handle invalid since the object wouldn't exist anymore! The properties of that object can be displayed using the **get** function. This shows properties such as the Color, LineStyle, LineWidth, and so on.

```
>> get(h1)
                  Color: [0 0 1]
              EraseMode: 'normal'
              LineStyle: '-'
              LineWidth: 0.5000
                 Marker: 'none'
             MarkerSize: 6
        MarkerEdgeColor: 'auto'
        MarkerFaceColor: 'none'
                  XData: [1x63 double]
                  YData: [1x63 double]
                  ZData: [1x0 double]
           BeingDeleted: 'off'
          ButtonDownFcn: []
```

```
          Children: [0x1 double]
          Clipping: 'on'
         CreateFcn: []
         DeleteFcn: []
        BusyAction: 'queue'
  HandleVisibility: 'on'
           HitTest: 'on'
     Interruptible: 'on'
          Selected: 'off'
SelectionHighlight: 'on'
               Tag: ' '
              Type: 'line'
     UIContextMenu: []
          UserData: []
           Visible: 'on'
            Parent: 157.5415
       DisplayName: ' '
         XDataMode: 'manual'
       XDataSource: ' '
       YDataSource: ' '
       ZDataSource: ' '
```

A particular property can also be displayed, for example, to see the line width:

```
>> get(h1,'LineWidth')
ans =
    0.5000
```

To see the objects, their properties, what the properties mean, and the valid values, go to the MATLAB Help. Under the Contents tab, click **Handle Graphics Property Browser**. Then, click **Plot Objects**; several options can be seen. Click **Lineseries**, which is used to create figures using the plot function, to see a list of the property names and a brief explanation of each.

For example, the Color property is a vector that stores the color of the line as three separate values for the red, green, and blue intensities, in that order. Each value is in the range from 0 (which means none of that color) to 1. In the previous example, the Color was [0 0 1], which means no red, no green, but full blue—in other words, the line drawn for the **sin** function was blue. Here are some more examples of possible values for the Color vector:

```
[1 0 0] is red
[0 1 0] is green
[0 0 1] is blue
[1 1 1] is white
[0 0 0] is black
[0.5 0.5 0.5] is a shade of grey
```

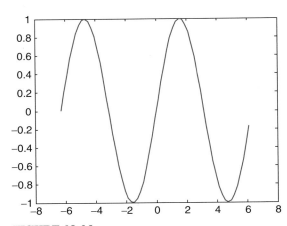

FIGURE 10.14

Plot of **sin** *function with increased line width.*

  PRACTICE 10.4

Create x and y vectors, and use the **plot** function to plot the data points represented by these vectors. Store the handle in a variable, and don't close the Figure Window! Use **get** to inspect the properties, and then **set** to change the line width and color.

All the properties listed by **get** can be changed, using the **set** function. The **set** function is called in the format

```
set(objhandle, 'PropertyName', property value)
```

For example, to change the line width from the default of 0.5 to 1.5:

```
>> set(hl,'LineWidth',1.5)
```

As long as the Figure Window is still open and this object handle is still valid, the width of the line will be increased.

The properties can also be set in the original function call. For example, this will increase the line width as seen in Figure 10.14.

```
>> hl = plot(x,y, 'Linewidth', 2.5);
```

10.6 PLOT APPLICATIONS

In this section, we will show some examples that integrate plots and many of the other concepts covered to this point in the book. For example, we will have a function that receives a *y* vector and a plot type as a string and will generate the plot, and we will show examples of reading data from a file and plotting it.

10.6.1 Plotting from a Function

The following function generates a Figure Window (seen in Figure 10.15) that shows different types of plots for the same *y* vector. The vector is passed as an input argument to the function, as is a cell array with the plot type names. The function generates the Figure Window using the cell array with the plot type names. It creates a function handle for each using the **str2func** function.

```
plotywithcell.m

function plotywithcell(y, rca)
% Shows different plot types, passed as a cell array
lenrca = length(rca);
for i = 1:lenrca
  subplot(1,lenrca,i)
```

(Continued)

```
   funh = str2func(rca{i});
   funh(y)
   title(upper(rca{i}))
 end
```

For example, the function could be
called as follows:

```
>> y = [1:2:9].^3
y =
    1   27   125   343   729
>> rca = {'bar', 'area', 'plot'};
>> plotywithcell(y, rca)
```

The function is general and works for
any number of plot types stored in the
cell array.

10.6.2 Plotting File Data

It is often necessary to read data from a
file and plot it. Normally, this entails
knowing the format of the file. For example, let us assume that a company has two
divisions, called A and B. The file 'ab06.
dat' has four lines in it (we will assume
this), with the sales figures (in millions) for the two divisions for each quarter of
the year 2006. For example, the file might look like this (the format will be exactly
like this):

```
A5.2B6.4
A3.2B5.5
A4.4B4.3
A4.5B2.2
```

FIGURE 10.15
*Subplot showing different
file types with their names
as titles.*

The following script reads the data and plots in one Figure Window the data as
bar charts. The script prints an error message if the file open is not successful or
if the file close was not successful. The **axis** command is used to force the x-axis
to range from 0 to 3 and the y-axis from 0 to 8, which will result in the axes
shown here. The numbers 1 and 2 would show on the x-axis rather than the
division labels A and B by default. The **set** function changes the XTickLabel
property to use the strings in the cell array as labels on the tick marks on the
x-axis; **gca** returns the handle to the axes in the current figure (it stands for
"get current axes").

plotdivab.m

```
% Reads sales figures for 2 divisions of a company one
% line at a time as strings, and plots the data
fid = fopen('ab06.dat');
if fid == -1
 disp('File open not successful')
else
 for i = 1:4
    % Every line is of the form A#B#; this separates
    % the characters and converts the #'s to actual
    % numbers
    aline = fgetl(fid);
    aline = aline(2:length(aline));
    [compa rest] = strtok(aline,'B');
    compa = str2num(compa);
    compb = rest(2:length(rest));
    compb = str2num(compb);
    % Data from every line is in a separate subplot
    subplot(1,4,i)
    bar([compa,compb])
    set(gca, 'XtickLabel', {'A', 'B'})
    axis([0 3 0 8])
    ylabel('Sales (millions)')
    title(sprintf('Quarter %d',i))
 end
 closeresult = fclose(fid);
 if closeresult ~= 0
    disp('File close not successful')
  end
end
```

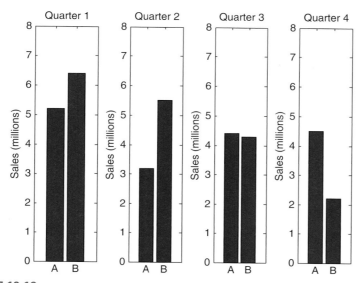

FIGURE 10.16

Subplot with customized x-axis tick labels.

FIGURE 10.17

Bar and pie charts with labels from file data.

Running this produces the subplot shown in Figure 10.16.

As another example, a data file called 'compsales.dat' stores sales figures (in millions) for divisions in a company. Each line in the file stores the sales number, followed by an abbreviation of the division name, in the following format:

```
5.2 X
3.3 A
5.8 P
2.9 Q
```

The following script uses the **textscan** function to read this information into a cell array, and then uses **subplot** to produce a Figure Window that displays the information in a **bar** chart and in a **pie** chart (see Figure 10.17).

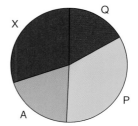

Sales in millions by division

compsalesbarpie.m

```matlab
% Reads sales figures and plots as a bar chart and a pie chart
fid = fopen('compsales.dat');
if fid == -1
 disp('File open not successful')
else
  % Use textscan to read the numbers and division codes
  % into separate elements in a cell array
  filecell = textscan(fid,'%f %s');
  % plot the bar chart with the division codes on the x ticks
  subplot(1,2,1)
  bar(filecell{1})
  xlabel('Division')
  ylabel('Sales (millions)')
  set(gca, 'XtickLabel', filecell{2})
  % plot the pie chart with the division codes as labels
  subplot(1,2,2)
  pie(filecell{1}, filecell{2})
  title('Sales in millions by division')
  closeresult = fclose(fid);
  if closeresult ~= 0
     disp('File close not successful'
   end
 end
```

SUMMARY

Common Pitfalls

- Forgetting that subplot numbers the plots rowwise rather than columnwise.

- Not realizing that the **subplot** function just creates a matrix within the Figure Window. Each part of this matrix must then be filled with a plot, using any type of plot function.

- Closing a Figure Window prematurely—the properties can be set only if the Figure Window is still open!

Programming Style Guidelines

- Always label plots.

- Take care to choose the type of plot in order to highlight the most relevant information.

MATLAB Functions and Commands			
subplot	hist	getframe	stem3
barh	pie	plot3	get
area	comet	bar3	set
stem	movie	pie3	

Exercises

1. Create a data file that has 10 numbers in it. Write a script that will load the vector from the file, and use **subplot** to do an **area** plot and a **stem** plot with this data in the same Figure Window. (**Note:** a loop is not needed.) Prompt the user for a title for each plot.

2. Use **subplot** to show the difference between the **sin** and **cos** functions. Create an x vector with 100 linearly spaced points in the range from -2π to 2π, and then two y vectors for **sin(x)** and **cos(x)**. In a 2×1 **subplot**, use the **plot** function to display them, with appropriate titles.

3. Biomedical engineers are developing an insulin pump for diabetics. To do this, it is important to understand how insulin is cleared from the body after a meal. The concentration of insulin at any time t is described by the equation

   $$C = C_0\ e^{-30t/m}$$

   where C_0 is the initial concentration of insulin, t is the time in minutes, and m is the mass of the person in kg. Write a script that will graphically show how the weight of the person influences the time for insulin to be cleared from the body. It will show in a 2×1 **subplot** the concentration of insulin for two subjects, one who weighs 120 pounds, and one who weighs 300 pounds. For both, the time should increment from 0 to 4 minutes in steps of 0.1 minute, and the initial concentration should be 85. The concentration over time will be shown in each subplot, and the weight of the person should be in the title. The conversion factor is 1 pound = 0.4536 kg. In order to better compare, use consistent axes for both plots.

4. Write a function *subfnfn* that will receive two function handles as input arguments, and will display in one Figure Window plots of these two functions, with the function names in the titles. Use the default axes. The function will create an x vector that ranges from 1 to n (where n is a random integer in the range from 4 to 10). For example, if the function is called as follows:

FIGURE 10.18
Subplot using function handles.

```
>> subfnfn(@sqrt, @exp)
```

and if the random integer for n was 9, the Figure Window would look like Figure 10.18.

5. A file called 'avehighs.dat' stores for 3 locations the average high temperatures for each month for a year (rounded to integers). There are three lines in the file; each stores the location number followed by the 12 temperatures (this may be assumed). For example, the file might store:

```
432 33   37   42   45   53   72   82   79   66   55   46   41
777 29   33   41   46   52   66   77   88   68   55   48   39
567 55   62   68   72   75   79   83   89   85   80   77   65
```

Write a script that will read this data and plot the temperatures for the three locations separately in one Figure Window. A **for** loop must be used to accomplish this. For example, if the data is as shown, the Figure Window would appear as Figure 10.19.

The axis labels and titles should be as shown.

6. Sales (in millions) from two different divisions of a company for the four quarters of 2006 are stored in vector variables, for example,

```
div1 = [4.2 3.8 3.7 3.8];
div2 = [2.5 2.7 3.1 3.3];
```

FIGURE 10.19

*Subplot to display data from file using a **for** loop.*

Using **subplot**, show side-by-side the sales figures for the two divisions. What kind of graph shows this in the best way? Why? In one graph, compare the two divisions. What kind of graph shows this in the best way? Why?

7. Create an x vector that has 30 linearly spaced points in the range from -2π to 2π, and then y as **sin(x)**. Do a **stem** plot of these points, and store the handle in a variable. Use **get** to see the properties of the stem plot, and then **set** to change the face color of the marker.

8. When an object with an initial temperature T is placed in a substance that has a temperature S, according to Newton's law of cooling, in t minutes it will reach a temperature T_t using the formula $T_t = S + (T - S) e^{(-kt)}$, where k is a constant value that depends on properties of the object. For an initial temperature of 100 and k = 0.6, graphically display the resulting temperatures from 1 to 10 minutes for two different surrounding temperatures: 50 and 20. Use the **plot** function to plot two different lines for these surrounding temperatures, and store the handle in a variable. Notice that two function handles are actually returned, and stored in a vector. Use **set** to change the line width of one of the lines.

9. Write a script that will draw the line y = x between x = 2 and x = 5, with a random thickness between 1 and 10.

10. In hydrology, ***hyetographs*** are used to display rainfall intensity during a storm. The intensity could be the amount of rain per hour, recorded every hour for a 24-hour period. Create your own data file to store the intensity in inches per hour every hour for 24 hours. Use a **bar** chart to display the intensities.

11. Write a script that will read x and y data points from a file, and will create an **area** plot with those points. The format of every line in the file is the letter x, a space, the x value, space, the letter y, space, and the y value. You must assume that the data file is in exactly that format, but you may not assume that the number of lines in the file is known. The number of points will be in the plot title. The script loops until the end of file is reached, using **fgetl** to read each line as a string. For example, IF the file contains the following

```
x 0 y 1
x 1.3 y 2.2
x 2.2 y 6
x 3.4 y 7.4
```

when running the script, the result will be as shown in Figure 10.20.

FIGURE 10.20
Area plot produced from x, y data read as strings from a file.

12. A file houseafford.dat stores on its three lines years, median incomes and median home prices for a city. The dollar amounts are in thousands. For example, it might look like this:

```
2000 2001 2002 2003 2004 2005 2006 2007
72 74 74 77 80 83 89 93
250 270 300 310 350 390 410 380
```

Create a file in this format, and then **load** the information into a matrix. Create a horizontal stacked bar chart to display the information shown as follows, with an appropriate title. **Note:** Use the XData property to put the years on the axis as shown in Figure 10.21.

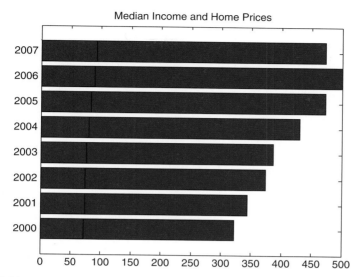

FIGURE 10.21
Horizontal stacked bar chart of median incomes and home prices.

13. A file 'houseafford.dat' stores on its three lines years, median incomes and median home prices for a city. The dollar amounts are in thousands. For example, it might look like this:

```
2000 2001 2002 2003 2004 2005 2006 2007
72 74 74 77 80 83 89 93
250 270 300 310 350 390 410 380
```

Create a file in this format, and then **load** the information into a matrix. The ratio of the home price to the income is called the housing affordability index. Calculate this for every year and plot it. The x-axis should show the years (e.g., 2000–2007). Store the handle of the plot in a variable and use **get** to see the properties and **set** to change at least one.

14. Do a quick survey of your friends to find out who prefers cheese pizza, pepperoni, or mushroom (no other possibilities; everyone must pick one of those three choices). Draw a pie chart to show the percentage favoring each. Label the pieces of this pizza pie chart!

15. The number of faculty members in each department at a certain College of Engineering is:

```
ME  22
BM  45
CE  23
EE  33
```

Experiment with at least three different plot types to graphically depict this information. Make sure that you have appropriate titles, labels, and legends on your plots. Which type(s) work best, and why?

16. The weights of the major components for a given aircraft are important considerations in aircraft design. The components include at the very least the wing, tail, fuselage, landing gear, and so on. Create a data file with values for these weights. Load the data from your file and create a pie chart to show the percentage weight for each component.

17. Create an *x* vector, and then two different vectors (*y* and *z*) based on *x*. Plot them with a legend. Use **help legend** to find out how to position the legend itself on the graph, and experiment with different locations.

18. The Wind Chill Factor (WCF) measures how cold it feels with a given air temperature T (in degrees Fahrenheit) and wind speed (V, in miles per hour). One formula for it is

$$WCF = 35.7 + 0.6T - 35.7 (V^{0.16}) + 0.43T (V^{0.16})$$

Experiment with different plot types to display the WCF for varying wind speeds and temperatures.

19. Write a script that will plot the **sin** function three times in one Figure Window, using the colors red, green, and blue.

20. Experiment with the **comet** function: Try the example given when **help comet** is entered and then animate your own function using **comet**.

21. Experiment with the **comet3** function: Try the example given when **help comet3** is entered and then animate your own function using **comet3**.

22. Investigate the **scatter** and **scatter3** functions.

23. The exponential and natural log functions are *inverse functions*. What does this mean in terms of the graphs of the functions? Show both functions in one Figure Window and distinguish between them.

24. The electricity generated by wind turbines annually in kilowatt-hours/year is given in a file. The amount of electricity is determined by, among other factors, the diameter of the turbine blade (in feet) and the wind velocity in mph. The file stores on each line the blade diameter, wind velocity, and the approximate electricity generated for the year. For example,

```
5  5  406
5  10  3250
5  15  10970
5  20  26000
10  5  1625
10  10  13000
```

```
10  15  43875
10  20  104005
20  5  6500
20  10  52000
20  15  175500
20  20  41600
```

Create this file, and determine how to graphically display this data.

CHAPTER 11

Solving Systems of Linear Algebraic Equations

KEY WORDS

linear algebraic
  equation
symbolic mathematics
matrix equality
square matrix
main diagonal
trace
symmetric
diagonal matrix
identity matrix
banded matrix
tridiagonal matrix

lower triangular matrix
upper triangular matrix
matrix addition
matrix subtraction
scalar multiplication
matrix multiplication
inner dimension
outer dimension
array multiplication
matrix inverse
matrix augmentation
dot product

cross product
coefficient
unknown
solution set
method of substitution
determinant
Gauss elimination
Gauss-Jordan elimination
elementary row
  operation
reduced row echelon
  form

CONTENTS

A *linear algebraic equation* is an equation of the form

$$a_1x_1 + a_2x_2 + a_3x_3 + \ldots + a_nx_n = b$$

Solutions to sets of equations in this form are important in many applications. In the MATLAB® product, to solve systems of equations, there are basically two methods:

321

- Using a matrix representation
- Using the **solve** function (which is part of Symbolic Math Toolbox™)

In this chapter, we will first investigate matrix and vector operations and then use these to solve linear algebraic equations. *Symbolic mathematics* means doing mathematics on symbols. Some of the symbolic math functions, which are in Symbolic Math Toolbox in MATLAB, will also be introduced, including the **solve** function for solving equations.

11.1 MATRIX DEFINITIONS

As we have already seen, a *matrix* can be thought of as a table of values in which there are both rows and columns. The general form of a matrix A (which is sometimes written as [A]) is shown here:

$$
A = \begin{bmatrix} a_{11} & a_{12} & \cdots & a_{1n} \\ a_{21} & a_{22} & \cdots & a_{2n} \\ \vdots & \vdots & \vdots & \vdots \\ a_{m1} & a_{m2} & \cdots & a_{mn} \end{bmatrix} = a_{ij} \quad i = 1, \ldots, m; \quad j = 1, \ldots, n
$$

This matrix has m rows and n columns, so the size is m × n.

A *vector* is a special case of a matrix, in which one of the dimensions (either m or n) is 1. A *row vector* is a 1 × n matrix. A *column vector* is an m × 1 matrix. A *scalar* is a special case of a matrix in which both m and n are 1, so it is a single value or a 1 × 1 matrix.

11.1.1 Matrix Properties

In this section we will define some special properties of matrices.

Two matrices are said to be *equal* to each other only if all corresponding elements are equal to each other. In order for this to be true, their sizes must be the same as well. The definition is [A] = [B] if $a_{ij} = b_{ij}$ for all i, j .

The Programming Concept

To test to see whether two matrices are equal to each other or not, both matrices are passed to a function that will return 1 for logical true if they are the same or 0 for logical false if not (or if they are not the same size). To write our own function, a flag is first set to 1, for logical true. If the two matrices are not the same size, the flag is set to 0 for false. Otherwise, using a nested **for** loop, each element in the first matrix argument *mata* is compared to the corresponding element in *matb*; if they are not the same, the flag is set to 0 for false.

myisequal.m

```
function myflag = myisequal(mata,matb)
% Assume that it's true that they are the same
% until and unless corresponding elements are
% found that are not the same
myflag = logical(1);
[r c] = size(mata);
if all(size(mata) ~= size(matb))
    myflag = logical(0);
else
  for i=1:r
    for j = 1:c
      if mata(i,j) ~= matb(i,j)
        myflag = logical(0);
      end
    end
  end
end
```

```
>> mata = [2 5 8; 1:3];
>> matb = [2:3:8; 1 2 3];
>> myisequal(mata,matb)
ans =
    1
```

The Efficient Method

Of course, in MATLAB, the function **all** could be used to check to see whether all elements are equal to each other, but in MATLAB, as we have seen, the **isequal** function will also accomplish this:

```
>> isequal(mata,matb)
ans =
    1
```

The **isequal** function will also return 0 for false if the two functions are not the same size.

11.1.2 Square Matrices

If a matrix has the same number of rows and columns, for example, if $m == n$, the matrix is *square*. The definitions that follow in this section apply only to square matrices.

The **main diagonal** of a square matrix is the set of terms a_{ii} for which the row and column indices are the same, for example, from the upper left element to the lower right. For example, for the following matrix it is the set of numbers 1, 6, 11, and 16.

$$\begin{bmatrix} 1 & 2 & 3 & 4 \\ 5 & 6 & 7 & 8 \\ 9 & 10 & 11 & 12 \\ 13 & 14 & 15 & 16 \end{bmatrix}$$

This is sometimes called just the **diagonal**. MATLAB has a function **diag** that will return the diagonal of a matrix as a column vector.

```
>> mymat = reshape(1:16,4,4)'
mymat =

    1     2     3     4
    5     6     7     8
    9    10    11    12
   13    14    15    16

>> diag(mymat)'
ans =

    1     6    11    16
```

The **trace** of a square matrix is the sum of all the elements on the diagonal. For example, for the preceding matrix it is $1 + 6 + 11 + 16$, or 34.

QUICK QUESTION!

How could we calculate the trace of a square matrix?

The Programming Concept

In order to calculate the trace of a square matrix, only one loop is necessary since the only elements in the matrix we're referring to are (i, i). So, once the size has been determined, the loop variable can iterate from 1 through the number of rows or from 1 through the number of columns (it doesn't matter which, since they have the same value!). The following function calculates and returns the trace of a square matrix, or an empty vector if the matrix argument is not square.

(Continued)

mytrace.m

```
function outsum = mytrace(mymat)
% Calculates the trace of a square matrix
% It returns an empty vector if the
% matrix is not square
[r c] = size(mymat);
if r ~= c
    outsum = [];
else
    outsum = 0;
    for i = 1:r
      outsum = outsum + mymat(i,i);
    end
end
```

```
>> mymat = reshape(1:16,4,4)'
mymat =

     1          2          3          4
     5          6          7          8
     9         10         11         12
    13         14         15         16

>> mytrace(mymat)
ans =
       34
```

The Efficient Method

In MATLAB, there is a built-in function **trace** to accomplish this:

```
>> trace(mymat)
ans =
       34
```

A square matrix is *symmetric* if $a_{ij} = a_{ji}$ for all i, j. In other words, all the values opposite the diagonal from each other must be equal to each other. In this example, there are three pairs of values opposite the diagonals, all of which are equal (the 2's, the 9's, and the 4's).

$$\begin{bmatrix} 1 & 2 & 9 \\ 2 & 5 & 4 \\ 9 & 4 & 6 \end{bmatrix}$$

A square matrix is a *diagonal matrix* if all values that are not on the diagonal are 0. The numbers on the diagonal, however, do not have to be all nonzero although frequently they are. Mathematically, this is written as $a_{ij} = 0$ for $i \sim = j$. Here is an example of a diagonal matrix.

$$\begin{bmatrix} 4 & 0 & 0 \\ 0 & 9 & 0 \\ 0 & 0 & 5 \end{bmatrix}$$

The Efficient Method

The **diag** function can be used to take a vector of length n and create an $n \times n$ square diagonal matrix with the values from the vector on the diagonal:

```
>> v=1:4;
>> diag(v)
ans =
```

1	0	0	0
0	2	0	0
0	0	3	0
0	0	0	4

A square matrix is an *identity matrix*, called [I], if $a_{ij} = 1$ for $i == j$ and $a_{ij} = 0$ for $i \sim = j$. In other words, all the numbers on the diagonal are 1 and all others are 0. This is a 3×3 identity matrix:

$$\begin{bmatrix} 1 & 0 & 0 \\ 0 & 1 & 0 \\ 0 & 0 & 1 \end{bmatrix}$$

Note that any identity matrix is a special case of a diagonal matrix.

MATLAB has a built-in function **eye** that will create an $n \times n$ identity matrix, given the value of n:

```
>> eye(5)
ans =
```

 PRACTICE 11.1

Write a function called **isdiagonal** that will return 1 for logical true if a square matrix is a diagonal matrix, or 0 for false if not.

$$\begin{bmatrix} 1 & 0 & 0 & 0 & 0 \\ 0 & 1 & 0 & 0 & 0 \\ 0 & 0 & 1 & 0 & 0 \\ 0 & 0 & 0 & 1 & 0 \\ 0 & 0 & 0 & 0 & 1 \end{bmatrix}$$

Note: i is built into MATLAB as the square root of -1, so another name is used for the function that creates an identity matrix: **eye**, which sounds like "i" (get it?).

There are several special cases of matrices that are related to diagonal matrices.

A *banded matrix* is a matrix of all 0's, with the exception of the main diagonal and other diagonals next to the main. For example, the following matrix has 0's except for the band of three diagonals; this is a particular kind of banded matrix called a *tridiagonal matrix*.

$$\begin{bmatrix} 1 & 2 & 0 & 0 \\ 5 & 6 & 7 & 0 \\ 0 & 10 & 11 & 12 \\ 0 & 0 & 15 & 16 \end{bmatrix}$$

A *lower triangular matrix* has all 0's above the main diagonal. For example,

$$\begin{bmatrix} 1 & 0 & 0 & 0 \\ 5 & 6 & 0 & 0 \\ 9 & 10 & 11 & 0 \\ 13 & 14 & 15 & 16 \end{bmatrix}$$

An *upper triangular matrix* has all 0's below the main diagonal. For example,

$$\begin{bmatrix} 1 & 2 & 3 & 4 \\ 0 & 6 & 7 & 8 \\ 0 & 0 & 11 & 12 \\ 0 & 0 & 0 & 16 \end{bmatrix}$$

It is possible for there to be 0's on the diagonal and in the lower part or upper part and still be a lower or upper triangular matrix, respectively.

The Efficient Method

MATLAB has functions **triu** and **tril** that will take a matrix and make it into an upper triangular or lower triangular matrix by replacing the appropriate elements with 0's.

```
>> mymat
mymat =

         1          2          3          4
         5          6          7          8
         9         10         11         12
        13         14         15         16

>> triu(mymat)
ans =

         1          2          3          4
         0          6          7          8
         0          0         11         12
         0          0          0         16

>> tril(mymat)
ans =

         1          0          0          0
         5          6          0          0
         9         10         11          0
        13         14         15         16
```

 PRACTICE 11.2

For the following matrices:

$$A \qquad B \qquad C$$

$$\begin{bmatrix} 4 & 3 \\ 0 & 1 \end{bmatrix} \quad \begin{bmatrix} 1 & 2 & 3 \\ 4 & 5 & 6 \end{bmatrix} \quad \begin{bmatrix} 1 & 4 & 2 \\ 4 & 0 & 3 \\ 2 & 3 & 6 \end{bmatrix}$$

Which are equal?

Which are square?

For all square matrices,

- Calculate the trace.
- Which are symmetric?
- Which are diagonal?
- Which are lower triangular?
- Which are upper triangular?

11.1.3 Matrix Operations

There are several common operations on matrices, some of which we have seen already. Operators that are applied term by term, implying that the matrices must be the same size, sometimes are referred to as *array operations*. These include addition and subtraction.

Matrix addition means adding two matrices term by term, which means they must be of the same size. In mathematical terms, this is written $c_{ij} = a_{ij} + b_{ij}$.

$$\begin{bmatrix} 1 & 2 & 3 \\ 4 & 5 & 6 \end{bmatrix} + \begin{bmatrix} 100 & 10 & 1 \\ 10 & 100 & 1 \end{bmatrix} = \begin{bmatrix} 101 & 12 & 4 \\ 14 & 105 & 7 \end{bmatrix}$$

The Programming Concept

In most languages, this would be accomplished using a nested **for** loop, as follows:

mymatadd.m

```
function outmat = mymatadd(mat1,mat2)
% Returns the sum of two matrices
% or [] if they are not the same size
outmat = [];
if all(size(mat1) == size(mat2))
  outmat = zeros(size(mat1)); % preallocate
  [r c] = size(mat1);
  for i = 1:r
    for j = 1:c
      outmat(i,j) = mat1(i,j) + mat2(i,j);
    end
  end
end
```

```
>> A = [1:3;4:6];
>> B = [100 10 1; 10 100 1];
>> C = mymatadd(A,B)
C =

    101         12          4
     14        105          7
```

The Efficient Method

However, in MATLAB this is accomplished with the + operator.

```
>> C = A + B
C =

    101         12          4
     14        105          7
```

Similar to matrix addition, **matrix subtraction** means to subtract term by term, so in mathematical terms $c_{ij} = a_{ij} - b_{ij}$. This would also be accomplished using a nested **for** loop in most languages, or by using the – operator in MATLAB.

Scalar multiplication means to multiply every element by a scalar (a number)

$$3 * \begin{bmatrix} 1 & 2 & 3 \\ 4 & 5 & 6 \end{bmatrix} = \begin{bmatrix} 3 & 6 & 9 \\ 12 & 15 & 18 \end{bmatrix}$$

This would also be accomplished using a nested for loop in most languages, or by using the * operator in MATLAB.

11.1.4 Matrix Multiplication

Matrix multiplication does *not* mean multiplying term by term; it is not an array operation. Matrix multiplication has a very specific meaning. First of all, in order to multiply a matrix A by a matrix B to result in a matrix C, the number of columns of A must be the same as the number of rows of B. If the matrix A has dimensions m × n, that means that matrix B must have dimensions n × *something*; we'll call it p. We say that the **inner dimensions** must be the same. The resulting matrix C has the same number of rows as A and the same number of columns as B (in other words, the **outer dimensions** m × p). In mathematical notation, $[A]_{m \times n} [B]_{n \times p} = [C]_{m \times p}$. This defines only the size of C.

The elements of the matrix C are defined as the sum of products of corresponding elements in the rows of A and columns of B, or in other words $c_{ij} = \sum_{k=1}^{n} a_{ik} b_{kj}$.

In the following example, A is 2 × 3 and B is 3 × 4 so C will be 2 × 4. The elements in C are obtained using the summation described earlier. The first row of C is obtained using the first row of A and in succession the columns of B. For example, C(1,1) is 3 * 1 + 8 * 4 + 0 * 0 or 35. C(1,2) is 3 * 2 + 8 * 5 + 0 * 2 or 46.

$$\begin{bmatrix} 3 & 8 & 0 \\ 1 & 2 & 5 \end{bmatrix} * \begin{bmatrix} 1 & 2 & 3 & 1 \\ 4 & 5 & 1 & 2 \\ 0 & 2 & 3 & 0 \end{bmatrix} = \begin{bmatrix} 35 & 46 & 17 & 19 \\ 9 & 22 & 20 & 5 \end{bmatrix}$$

The Programming Concept

In order to multiply two matrices together, three nested **for** loops are required. The two outer loops iterate through the rows and columns of C, which is m × p. For each element in C, the inner loop sums $a_{ik} * b_{kj}$ for values of k from 1 through n. Here is a script that will accomplish this:

mymatmult.m

```
% This script demonstrates matrix multiplication
A = [3 8 0; 1 2 5];
B = [1 2 3 1; 4 5 1 2; 0 2 3 0];
[m n] = size(A);
```

(Continued)

```
 [nb p] = size(B);
 if n ~= nb
    disp('Cannot perform this matrix multiplication')
 else
    % Outer 2 loops iterate through the elements in C
    % which has dimensions m by p
    % Preallocate C
    C = zeros(m,p);
    for i=1:m
      for j = 1:p
        % Inner loop performs the sum for each
        % element in C
        mysum = 0;
        for k = 1:n
          mysum = mysum + A(i,k) * B(k,j);
        end
        C(i,j) = mysum;
      end
    end
    C
 end
end
```

```
>> mymatmult
C =

   35        46        17        19
    9        22        20         5
```

The Efficient Method

In MATLAB, the * operator will perform this matrix multiplication:

```
>> A = [3 8 0; 1 2 5];
>> B = [1 2 3 1; 4 5 1 2; 0 2 3 0];
>> C = A*B
C =

   35        46        17        19
    9        22        20         5
```

QUICK QUESTION!

What happens if a matrix M is multiplied by an identity matrix (of the appropriate size)?

Answer: For the size to be appropriate, the identity matrix would have the same dimensions as the number of columns of M. The result of the multiplication will always be the original matrix M (so, it is similar to multiplying a scalar by 1).

```
>> A = [3 8 0; 1 2 5]
A =

      3        8        0
      1        2        5
>> A * eye(3)
ans =

      3        8        0
      1        2        5
>> B = [1 2 3 1; 4 5 1 2; 0 2 3 0]
B =

      1        2        3        1
      4        5        1        2
      0        2        3        0
>> [r c] = size(B);
>> B * eye(c)
ans =

      1        2        3        1
      4        5        1        2
      0        2        3        0
```

To multiply matrices term by term (which again is *not* called matrix multiplication!) in MATLAB, the .* operator is used, and the matrices must have the same dimensions. This is called *array multiplication* since it is an array operation (term by term).

PRACTICE 11.3

Multiply these two matrices by hand and then verify the result in MATLAB.

$$\begin{matrix} A & B \end{matrix}$$

$$\begin{bmatrix} 1 & 3 \\ 4 & 2 \end{bmatrix} \quad \begin{bmatrix} 2 & 1 \\ -1 & 3 \end{bmatrix}$$

```
>> A = [1:3;4:6];
>> B = [100 10 1; 10 100 1];
>> C = A .* B
C =

      100          20          3
       40         500          6
```

A *matrix transpose* interchanges the rows and columns of a matrix. For a matrix A, its transpose is written A^T. For example, if $A = \begin{bmatrix} 1 & 2 & 3 \\ 4 & 5 & 6 \end{bmatrix}$ then $A^T = \begin{bmatrix} 1 & 4 \\ 2 & 5 \\ 3 & 6 \end{bmatrix}$.

In MATLAB, as we have seen, there is a built-in transpose operator, the apostrophe.

If the result of multiplying a matrix A by a matrix B is the identity matrix, then B is the *inverse* of matrix A. The inverse of a matrix A is written as A^{-1}, so [A][A^{-1}] = [I]. How to actually compute the inverse A^{-1} of a matrix by hand is not so easy. MATLAB, however, has a function **inv** to compute a matrix inverse. For example, here a matrix is created, its inverse is found, and then multiplied by the original matrix to verify that the product is in fact the identity matrix:

```
>> a = [1 2; 2 2]
a =

     1           2
     2           2

>> ainv = inv(a)
ainv =

    -1.0000      1.0000
     1.0000     -0.5000

>> a*ainv
ans =

     1           0
     0           1
```

Matrix augmentation means adding column(s) to the original matrix. For example, the matrix A

$$A = \begin{bmatrix} 1 & 3 & 7 \\ 2 & 5 & 4 \\ 9 & 8 & 6 \end{bmatrix}$$

might be augmented with a 3×3 identity matrix:

$$\begin{bmatrix} 1 & 3 & 7 & 1 & 0 & 0 \\ 2 & 5 & 4 & 0 & 1 & 0 \\ 9 & 8 & 6 & 0 & 0 & 1 \end{bmatrix}$$

Sometimes in mathematics the vertical line is shown to indicate that the matrix has been augmented. In MATLAB, matrix augmentation can be accomplished using square brackets to concatenate the two matrices. The square matrix a is concatenated with an identity matrix that has the same size as the matrix a:

```
>> a = [1 3 7; 2 5 4; 9 8 6]
a =

    1       3       7
    2       5       4
    9       8       6
>> [a eye(size(a))]
ans =

    1       3       7       1       0       0
    2       5       4       0       1       0
    9       8       6       0       0       1
```

Of course, as we have seen already, it is more efficient to preallocate the matrix to the correct dimensions to begin with. Particularly for large matrices, augmenting a matrix in this fashion is inefficient.

11.1.5 Vector Operations

Since vectors are just special cases of matrices, the matrix operations described earlier (addition, subtraction, scalar multiplication, multiplication, transpose) work on vectors as well, as long as the dimensions are correct.

For vectors, we have already seen that the transpose of a row vector is a column vector, and the transpose of a column vector is a row vector.

To multiply vectors, they must have the same number of elements, but one must be a row vector and the other a column vector. For example, for a column vector c and row vector r:

$$c = \begin{bmatrix} 5 \\ 3 \\ 7 \\ 1 \end{bmatrix} \quad r = \begin{bmatrix} 6 & 2 & 3 & 4 \end{bmatrix}$$

Note that r is a 1×4, and c is 4×1. Recall that to multiply two matrices,

$$[A]_{m \times n}[B]_{n \times p} = [C]_{m \times p}$$

so $[r]_{1 \times 4} [c]_{4 \times 1} = [s]_{1 \times 1}$, or in other words a scalar:

$$\begin{bmatrix} 6 & 2 & 3 & 4 \end{bmatrix} \begin{bmatrix} 5 \\ 3 \\ 7 \\ 1 \end{bmatrix} = 6*5 + 2*3 + 3*7 + 4*1 = 61$$

whereas $[c]_{4 \times 1} [r]_{1 \times 4} = [M]_{4 \times 4}$, or in other words a 4×4 matrix:

$$\begin{bmatrix} 5 \\ 3 \\ 7 \\ 1 \end{bmatrix} \begin{bmatrix} 6 & 2 & 3 & 4 \end{bmatrix} = \begin{bmatrix} 30 & 10 & 15 & 20 \\ 18 & 6 & 9 & 12 \\ 42 & 14 & 21 & 28 \\ 6 & 2 & 3 & 4 \end{bmatrix}$$

In MATLAB, these operations are accomplished using the * operator, which is the matrix multiplication operator. First, the column vector c and row vector r are created.

```
>> c = [5 3 7 1]';
>> r = [6 2 3 4];
>> r*c
ans =
      61
>> c*r
ans =
      30        10        15        20
      18         6         9        12
      42        14        21        28
       6         2         3         4
```

11.1.5.1 Dot Product

The *dot product* or *inner product* of two vectors a and b is written as a • b and is defined as as $a_1 b_1 + a_2 b_2 + a_3 b_3 + \dots + a_n b_n = \sum_{i=1}^{n} a_i b_i$.

In other words, this is like matrix multiplication when multiplying a row vector a by a column vector b; the result is a scalar. This can be accomplished using the * operator and transposing the second vector, or by using the **dot** function in MATLAB:

```
>> vec1 = [4 2 5 1];
>> vec2 = [3 6 1 2];
>> vec1*vec2'
ans =
      31
>> dot(vec1,vec2)
ans =
      31
```

11.1.5.2 Cross Product

The *cross product* or *outer product* a × b of two vectors a and b is defined only when both a and b are vectors in three-dimensional space, which means that they both must have three elements. It can be defined as a matrix multiplication of a matrix composed from the elements from a in a particular manner shown here and the column vector b.

$$a \times b = \begin{bmatrix} 0 & -a_3 & a_2 \\ a_3 & 0 & -a_1 \\ -a_2 & a_1 & 0 \end{bmatrix} \begin{bmatrix} b_1 \\ b_2 \\ b_3 \end{bmatrix} = \begin{bmatrix} a_2 b_3 - a_3 b_2, & a_3 b_1 - a_1 b_2 - a_2 b_1 \end{bmatrix}$$

MATLAB has a built-in function **cross** to accomplish this.

```
>> vec1 = [4 2 5 ];
>> vec2 = [3 6 1 ];
>> cross(vec1,vec2)
ans =
    -28      11      18
```

11.2 MATRIX SOLUTIONS TO SYSTEMS OF LINEAR ALGEBRAIC EQUATIONS

A *linear algebraic equation* is an equation of the form

$$a_1 x_1 + a_2 x_2 + a_3 x_3 + \dots + a_n x_n = b$$

where the a's are constant *coefficients*, the x's are the **unknowns**, and b is a *constant*. A solution is a sequence of numbers s_1, s_2, and s_3 that satisfy the equation. For example,

$$4x_1 + 5x_2 - 2x_3 = 16$$

is such an equation in which there are three unknowns: x_1, x_2, and x_3. One solution to this equation is $x_1 = 3$, $x_2 = 4$, and $x_3 = 8$, since $4 * 3 + 5 * 4 - 2 * 8$ is equal to 16.

A *system* of linear algebraic equations is a set of equations of the form:

$$a_{11} x_1 + a_{12} x_2 + a_{13} x_3 + \dots + a_{1n} x_n = b_1$$

$$a_{21} x_1 + a_{22} x_2 + a_{23} x_3 + \dots + a_{2n} x_n = b_2$$

$$a_{31} x_1 + a_{32} x_2 + a_{33} x_3 + \dots + a_{3n} x_n = b_3$$

$$\vdots \qquad \vdots \qquad \vdots \qquad \vdots \qquad \vdots$$

$$a_{m1} x_1 + a_{m2} x_2 + a_{m3} x_3 + \dots + a_{mn} x_n = b_m$$

This is called an m × n system of equations; there are m equations and n unknowns.

Because of the way that matrix multiplication works, these equations can be represented in matrix form as Ax = b where A is a matrix of the coefficients, x is a column vector of the unknowns, and b is a column vector of the constants from the right-hand side of the equations:

$$
A \qquad\qquad x \quad = \quad b
$$

$$
\begin{bmatrix}
a_{11} & a_{12} & a_{13} & \cdots & a_{1n} \\
a_{12} & a_{22} & a_{23} & \cdots & a_{2n} \\
a_{31} & a_{32} & a_{33} & \cdots & a_{3n} \\
\vdots & \vdots & \vdots & \vdots & \vdots \\
a_{m1} & a_{m2} & a_{m3} & \cdots & a_{mn}
\end{bmatrix}
\begin{bmatrix}
x_1 \\ x_2 \\ x_3 \\ \vdots \\ x_n
\end{bmatrix}
=
\begin{bmatrix}
b_1 \\ b_2 \\ b_3 \\ \vdots \\ b_m
\end{bmatrix}
$$

A *solution set* is the set of all possible solutions to the system of equations (all sets of values for the unknowns that solve the equations). All systems of linear equations have either:

- No solutions
- One solution
- Infinitely many solutions

One of the main concepts of the subject of *linear algebra* is the different methods of solving (or attempting to solve!) systems of linear algebraic equations. MATLAB has many functions that assist in this process.

Once the system of equations has been written in matrix form, what we want is to solve the equation Ax = b for the unknown x. To do this, we need to isolate x on one side of the equation. If we were working with scalars, we would divide both sides of the equation by x. In fact, with MATLAB we can use the **divided into** operator to do this. However, most languages cannot do this with matrices, so we instead multiply both sides of the equation by the inverse of the coefficient matrix A:

```
A⁻¹ A x = A⁻¹ b
```

Then, because multiplying a matrix by its inverse results in the identity matrix I, and because multiplying any matrix by I results in the original matrix, we have:

```
I x = A⁻¹ b
```

or

```
x = A⁻¹ b
```

This means that the column vector of unknown x is found as the inverse of matrix A multiplied by the column vector b. So, if we can find the inverse of A, we can solve for the unknown in x.

For example, consider the following three equations with three unknowns x_1, x_2, and x_3:

```
 4x₁ - 2x₂ + 1x₃ = 7
 1x₁ + 1x₂ + 5x₃ = 10
-2x₁ + 3x₂ - 1x₃ = 2
```

We write this in the form $Ax = b$ where A is a matrix of the coefficients, x is a column vector of the unknown x_i, and b is a column vector of the values on the right-hand side of the equations:

$$
\begin{array}{ccc}
A & x & b
\end{array}
$$

$$
\begin{bmatrix} 4 & -2 & 1 \\ 1 & 1 & 5 \\ -2 & 3 & -1 \end{bmatrix}
\begin{bmatrix} x_1 \\ x_2 \\ x_3 \end{bmatrix}
=
\begin{bmatrix} 7 \\ 10 \\ 2 \end{bmatrix}
$$

The solution is then $x = A^{-1} b$. In MATLAB there are two simple ways to solve this. The built-in function **inv** can be used to get the inverse of A and then multiply this by b, or we can use the **divided into** operator.

```
>> A = [4 -2 1; 1 1 5; -2 3 -1];
>> b = [7;10;2];
>> x = inv(A)*b
x =
    3.0244
    2.9512
    0.8049
>> x = A\b
x =
    3.0244
    2.9512
    0.8049
```

11.2.1 Solving 2 × 2 Systems of Equations

Although this may seem easy in MATLAB, in general finding solutions to systems of equations is not. Systems that are 2 × 2 are, however, fairly straightforward, and there are several methods of solution for these systems for which MATLAB has built-in functions.

Consider the following 2 × 2 system of equations:

$$
\begin{aligned}
x_1 + 2x_2 &= 2 \\
2x_1 + 2x_2 &= 6
\end{aligned}
$$

First, to visualize the solution, it will be easier to change both equations to the equation of a straight line by writing each in the form $y = mx + b$ (by changing x_1 to x and x_2 to y):

$$
\begin{array}{ccccc}
x + 2y = 2 & \rightarrow & 2y = -x + 2 & \rightarrow & y = -0.5x + 1 \\
2x + 2y = 6 & \rightarrow & 2y = -2x + 6 & \rightarrow & y = -x + 3
\end{array}
$$

In MATLAB we can plot these lines using a script; the results are seen in Figure 11.1.

plot2by2.m

```
% Plot a 2 by 2 system as straight lines
x = -2:5;
y1 = -0.5 * x + 1;
y2 = -x + 3;
plot(x,y1,x,y2)
axis([-2 5 -4 6])
```

The intersection of the lines is the point $(4, -1)$. In other words, $x = 4$ and $y = -1$. Changing back to x_1 and x_2, we have $x_1 = 4$ and $x_2 = -1$. This allows us to visualize the solution.

This system of equations in matrix form is:

$$\underset{A}{\begin{bmatrix} 1 & 2 \\ 2 & 2 \end{bmatrix}} \underset{x}{\begin{bmatrix} x_1 \\ x_2 \end{bmatrix}} = \underset{b}{\begin{bmatrix} 2 \\ 6 \end{bmatrix}}$$

We have already seen that the solution is $x = A^{-1} b$, so we can solve this if we can find the inverse of A. One method of finding the inverse for a 2×2 matrix involves calculating the *determinant* D.

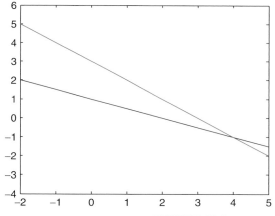

FIGURE 11.1

Visualizing 2 × 2 system of equations as straight lines.

For a 2×2 matrix $A = \begin{bmatrix} a_{11} & a_{12} \\ a_{21} & a_{22} \end{bmatrix}$, the determinant D is defined as:

$$D = \begin{vmatrix} a_{11} & a_{12} \\ a_{21} & a_{22} \end{vmatrix} = a_{11} a_{22} - a_{12} a_{21}$$

It is written using vertical lines around the coefficients of the matrix, and is defined as the product of the values on the diagonal minus the product of the other two numbers.

For a 2×2 matrix, the matrix inverse is defined in terms of D as

$$A^{-1} = \frac{1}{D} \begin{bmatrix} a_{22} & -a_{12} \\ -a_{21} & a_{11} \end{bmatrix}$$

The inverse is therefore the result of multiplying the scalar $1/D$ by every element in the preceding matrix. Notice that this is not the matrix A, but is determined using the elements from A in the following manner: the values on the diagonal are reversed, and the negation operator is used on the other two values.

Notice that if the determinant D is 0, it will not be possible to find the inverse of the matrix A.

For our coefficient matrix:

$$A = \begin{bmatrix} 1 & 2 \\ 2 & 2 \end{bmatrix}, D = \begin{vmatrix} 1 & 2 \\ 2 & 2 \end{vmatrix} = 1*2 - 2*2 \text{ or } -2$$

so.

$$A^{-1} = \frac{1}{1*2-2*2} \begin{bmatrix} 2 & -2 \\ -2 & 1 \end{bmatrix} = \frac{1}{-2} \begin{bmatrix} 2 & -2 \\ -2 & 1 \end{bmatrix} = \begin{bmatrix} -1 & 1 \\ 1 & -\frac{1}{2} \end{bmatrix}$$

Therefore,

$$\begin{bmatrix} x_1 \\ x_2 \end{bmatrix} = \begin{bmatrix} -1 & 1 \\ 1 & -\frac{1}{2} \end{bmatrix} \begin{bmatrix} 2 \\ 6 \end{bmatrix}$$

The unknowns are found by performing this matrix multiplication, so:

```
x₁ = -1 * 2 + 1 * 6 = 4
x₂ = 1 * 2 + (-1/2) * 6 = -1
```

This, of course, is the same solution as found by the intersection of the two lines.

To do this in MATLAB, first we would create the coefficient matrix variable *a* and column vector *b*.

```
>> a = [1 2; 2 2];
>> b = [2;6];
```

The Programming Method

For 2×2 matrices, the determinant and inverse are found using simple expressions.

```
>> deta = a(1,1)*a(2,2) - a(1,2)*a(2,1)
deta =
      -2
>> inva = (1/deta) * [a(2,2) -a(1,2); -a(2,1) a(1,1)]
       inva =

   -1.0000          1.0000
    1.0000         -0.5000
```

 PRACTICE 11.4

For the following 2×2 system of equations:

```
x₁ + 3x₂ = 2
2x₁ + 4x₂ = 5
```

Do the following on paper:

- Write the equations in matrix form Ax = b.
- Solve by finding the inverse A⁻¹ and then x = A⁻¹ b.

Now, get into MATLAB and check them.

The Efficient Method

We have already seen that MATLAB has a built-in function, **inv**, to find a matrix inverse. It also has a built-in function **det** to find a determinant:

```
>> det(a)
ans =
      -2
>> inv(a)
ans =
```

$$
\begin{array}{cc}
-1.0000 & 1.0000 \\
1.0000 & -0.5000
\end{array}
$$

11.2.2 Gauss, Gauss-Jordan Elimination

For 2×2 systems of equations, there are well-defined, simple solution methods. However, for larger systems of equations, finding solutions is frequently not as straightforward.

Two related methods of solving systems of linear equations will be described here: *Gauss elimination*, and *Gauss-Jordan elimination*. They are both based on the observation that systems of equations are equivalent if they have the same solution set. Also, performing simple operations on the rows of a matrix, called Elementary Row Operations (EROs), result in equivalent systems. These fall into three categories:

1. Replacement: replace a row by adding it to (or subtracting from it) a multiple of another row. For a given row r_i, this is written as

$$r_i \leftarrow sr_j \rightarrow r_i$$

Note that when replacing row r_i, nothing is multiplied by it. Instead, row r_j is multiplied by a scalar s (which could be a fraction) and that is added to or subtracted from row r_i.

2. Interchange rows: for example interchanging rows r_i and r_j is written as

$$r_i \longleftrightarrow r_j$$

3. Scaling: changes a row by multiplying it by a nonzero scalar $sr_i \rightarrow r_i$

For example, for the matrix:

$$
\begin{bmatrix}
4 & 2 & 3 \\
1 & 4 & 0 \\
2 & 5 & 3
\end{bmatrix}
$$

An example of interchanging rows would be $r_1 \longleftrightarrow r_3$, which would yield:

$$
\begin{bmatrix}
4 & 2 & 3 \\
1 & 4 & 0 \\
2 & 5 & 3
\end{bmatrix}
\quad r_1 \longleftrightarrow r_3 \quad
\begin{bmatrix}
2 & 5 & 3 \\
1 & 4 & 0 \\
4 & 2 & 3
\end{bmatrix}
$$

Now, starting with this matrix, an example of scaling would be: $2r_2 \rightarrow r_2$, which means all elements in row 2 are multiplied by 2. This yields:

$$
\begin{bmatrix}
2 & 5 & 3 \\
1 & 4 & 0 \\
4 & 2 & 3
\end{bmatrix}
\quad 2r_2 \rightarrow r_2 \quad
\begin{bmatrix}
2 & 5 & 3 \\
2 & 8 & 0 \\
4 & 2 & 3
\end{bmatrix}
$$

Now, starting with this matrix, an example of a replacement would be: $r_3 - 2r_2 \rightarrow r_3$. Element-by-element, row 3 is replaced by the element in row 3 minus 2 * the corresponding element in row 2. This yields:

$$\begin{bmatrix} 2 & 5 & 3 \\ 2 & 8 & 0 \\ 4 & 2 & 3 \end{bmatrix} r_3 - 2r_2 \rightarrow r_3 \begin{bmatrix} 2 & 5 & 3 \\ 2 & 8 & 0 \\ 0 & -14 & 3 \end{bmatrix}$$

 PRACTICE 11.5

Show the result of each of the following EROs:

$$\begin{bmatrix} 4 & 2 & 3 \\ 1 & 4 & 0 \\ 2 & 5 & 3 \end{bmatrix} r_1 \longleftrightarrow r_2$$

$$\begin{bmatrix} 4 & 2 & 3 \\ 1 & 4 & 0 \\ 2 & 5 & 3 \end{bmatrix} r_2 - \frac{1}{4} r_1 \rightarrow r_2$$

$$\begin{bmatrix} 4 & 2 & 3 \\ 1 & 4 & 0 \\ 2 & 5 & 3 \end{bmatrix} \frac{1}{2} r_1 \rightarrow r_1$$

Both the Gauss and Gauss-Jordan methods begin with the matrix form Ax = b of a system of equations, and then augment the coefficient matrix A with the column vector b.

11.2.2.1 Gauss Elimination

The Gauss elimination method consists of:

- Creating the augmented matrix [A b]
- Applying EROs to this augmented matrix to get an upper triangular form (this is called *forward elimination*)
- *Back-substitution* to solve

For example, for a 2 × 2 system, the augmented matrix would be:

$$\begin{bmatrix} a_{11} & a_{12} & b_1 \\ a_{21} & a_{22} & b_2 \end{bmatrix}$$

Then, EROs are applied to get the augmented matrix into an upper triangular form:

$$\begin{bmatrix} a_{11}' & a_{12}' & b_1' \\ 0 & a_{22}' & b_2' \end{bmatrix}$$

So, the goal is simply to replace a_{21} with 0. Here, the primes indicate that the values (may) have been changed.

Putting this back into the equation form yields

$$\begin{bmatrix} a_{11}' & a_{12}' \\ 0 & a_{22}' \end{bmatrix} \begin{bmatrix} x_1 \\ x_2 \end{bmatrix} = \begin{bmatrix} b_1' \\ b_2' \end{bmatrix}$$

Performing this matrix multiplication for each row results in:

$a'_{11} x_1 + a'_{12} x_2 = b'_1$
$a'_{22} x_2 = b'_2$

So, the solution is:

$x_2 = b'_2 / a'_{22}$
$x_1 = (b'_1 - a'_{12} x_2) / a'_{11}$

Similarly, for a 3×3 system, the augmented matrix is reduced to upper triangular form:

$$\begin{bmatrix} a_{11} & a_{12} & a_{13} & b_1 \\ a_{21} & a_{22} & a_{23} & b_2 \\ a_{31} & a_{32} & a_{33} & b_3 \end{bmatrix} \rightarrow \begin{bmatrix} a'_{11} & a'_{12} & a'_{13} & b'_1 \\ 0 & a'_{22} & a'_{23} & b'_2 \\ 0 & 0 & a'_{33} & b'_3 \end{bmatrix}$$

(This will be done systematically by first getting a 0 in the a_{21} position, then a_{31}, and finally a_{32}.)

Then, the solution will be:

$x_3 = b_3' / a_{33}'$
$x_2 = (b_2' - a_{23}'x_3) / a_{22}'$
$x_1 = (b_1' - a_{13}'x_3 - a_{12}'x_2) / a_{11}'$

As an example, consider the following 2×2 system of equations:

$x_1 + 2x_2 = 2$
$2x_1 + 2x_2 = 6$

As a matrix equation $Ax = b$, this is:

$$\begin{bmatrix} 1 & 2 \\ 2 & 2 \end{bmatrix} \begin{bmatrix} x_1 \\ x_2 \end{bmatrix} = \begin{bmatrix} 2 \\ 6 \end{bmatrix}$$

The first step is to augment the coefficient matrix A with b to get an augmented matrix [A| b]:

$$\begin{bmatrix} 1 & 2 & 2 \\ 2 & 2 & 6 \end{bmatrix}$$

For forward elimination, we want to get a 0 in the a_{21} position. To accomplish this, we can modify the second line in the matrix by subtracting from it 2 * the first row.

The way we would write this ERO is:

$$\begin{bmatrix} 1 & 2 & 2 \\ 2 & 2 & 6 \end{bmatrix} r_2 - 2r_1 \rightarrow r_2 \begin{bmatrix} 1 & 2 & 2 \\ 0 & -2 & 2 \end{bmatrix}$$

Now, putting it back in matrix equation form:

$$\begin{bmatrix} 1 & 2 \\ 0 & -2 \end{bmatrix} \begin{bmatrix} x_1 \\ x_2 \end{bmatrix} = \begin{bmatrix} 2 \\ 2 \end{bmatrix}$$

says that the second equation is now $-2x_2 = 2$ so $x_2 = -1$. Plugging into the first equation:

$x_1 + 2(-1) = 2$, so $x_1 = 4$.

This is called back-substitution.

11.2.2.2 Gauss-Jordan

The Gauss-Jordan elimination method starts the same way that the Gauss elimination method does, but then instead of back-substitution, the elimination continues. The Gauss-Jordan method consists of:

- Creating the augmented matrix [A b]
- Forward elimination by applying EROs to get an upper triangular form
- Back elimination to a diagonal form which yields the solution

For a 2×2 system, this would yield

$$\begin{bmatrix} a_{11} & a_{12} & b_1 \\ a_{21} & a_{22} & b_2 \end{bmatrix} \rightarrow \begin{bmatrix} a'_{11} & 0 & b'_1 \\ 0 & a'_{22} & b'_2 \end{bmatrix}$$

and for a 3×3 system,

$$\begin{bmatrix} a_{11} & a_{12} & a_{13} & b_1 \\ a_{21} & a_{22} & a_{23} & b_2 \\ a_{31} & a_{32} & a_{33} & b_3 \end{bmatrix} \rightarrow \begin{bmatrix} a'_{11} & 0 & 0 & b'_1 \\ 0 & a'_{22} & 0 & b'_2 \\ 0 & 0 & a'_{33} & b'_3 \end{bmatrix}$$

Notice that the resulting diagonal form does not include the right-most column.

For example, for the 2×2 system, forward elimination yielded the matrix:

$$\begin{bmatrix} 1 & 2 & 2 \\ 0 & -2 & 2 \end{bmatrix}$$

Now, to continue with back elimination, we need a 0 in the a_{12} position.

$$\begin{bmatrix} 1 & 2 & 2 \\ 0 & -2 & 2 \end{bmatrix} \xrightarrow{r_1 + r_2 \rightarrow r_1} \begin{bmatrix} 1 & 0 & 4 \\ 0 & -2 & 2 \end{bmatrix}$$

So, the solution is $x_1 = 4$; $-2x_2 = 2$ or $x_2 = -1$.

Here is an example of a 3×3 system:

$$\begin{aligned} x_1 + 3x_2 \quad\quad &= 1 \\ 2x_1 + x_2 + 3x_3 &= 6 \\ 4x_1 + 2x_2 + 3x_3 &= 3 \end{aligned}$$

In matrix form, the augmented matrix $[A|b]$ is

$$\left[\begin{array}{ccc|c} 1 & 3 & 0 & 1 \\ 2 & 1 & 3 & 6 \\ 4 & 2 & 3 & 3 \end{array}\right]$$

Forward substitution (done systematically by first getting a 0 in the a_{21} position, then a_{31}, and finally a_{32}):

$$\left[\begin{array}{ccc|c} 1 & 3 & 0 & 1 \\ 2 & 1 & 3 & 6 \\ 4 & 2 & 3 & 3 \end{array}\right] r_2 - 2r_1 \rightarrow r_2 \left[\begin{array}{ccc|c} 1 & 3 & 0 & 1 \\ 0 & -5 & 3 & 4 \\ 4 & 2 & 3 & 3 \end{array}\right] r_3 - 4r_1 \rightarrow r_3 \left[\begin{array}{ccc|c} 1 & 3 & 0 & 1 \\ 0 & -5 & 3 & 4 \\ 0 & -10 & 3 & -1 \end{array}\right]$$

$$r_3 - 2r_2 \rightarrow r_3 \left[\begin{array}{ccc|c} 1 & 3 & 0 & 1 \\ 0 & -5 & 3 & 4 \\ 0 & 0 & -3 & -9 \end{array}\right]$$

For the Gauss method, this is followed by back-substitution:

$$\begin{aligned} -3x_3 &= -9 \\ x_3 &= 3 \\ -5x_2 + 3(3) &= 4 \\ -5x_2 &= -5 \\ x_2 &= 1 \\ x_1 + 3(1) &= 1 \\ x_1 &= -2 \end{aligned}$$

For the Gauss-Jordan method, this is instead followed by back elimination:

$$\left[\begin{array}{ccc|c} 1 & 3 & 0 & 1 \\ 0 & -5 & 3 & 4 \\ 0 & 0 & -3 & -9 \end{array}\right] r_2 - r_3 \rightarrow r_2 \left[\begin{array}{ccc|c} 1 & 3 & 0 & 1 \\ 0 & -5 & 0 & -5 \\ 0 & 0 & -3 & -9 \end{array}\right]$$

$$r_1 + \tfrac{3}{5} r_2 \rightarrow r_1 \left[\begin{array}{ccc|c} 1 & 0 & 0 & -2 \\ 0 & -5 & 0 & -5 \\ 0 & 0 & -3 & -9 \end{array}\right]$$

So

$$x_1 = -2$$
$$-5x_2 = -5$$
$$x_2 = 1$$
$$-3x_3 = -9$$
$$x_3 = 3$$

Here's an example of performing these substitutions using MATLAB:

```
>> a = [1 3 0; 2 1 3; 4 2 3]
a =

    1      3      0
    2      1      3
    4      2      3
>> b = [1 6 3]'
b =

    1
    6
    3
>> ab = [a b]
ab =

    1      3      0      1
    2      1      3      6
    4      2      3      3
>> ab(2,:) = ab(2,:) - 2*ab(1,:)
ab =

    1      3      0      1
    0     -5      3      4
    4      2      3      3
>> ab(3,:) = ab(3,:) - 4 * ab(1,:)
ab =

    1      3      0      1
    0     -5      3      4
    0    -10      3     -1
>> ab(3,:) = ab(3,:) - 2 * ab(2,:)
ab =

    1      3      0      1
    0     -5      3      4
    0      0     -3     -9
```

```
>> ab(2,:) = ab(2,:) + ab(3,:)
ab =

    1       3       0       1
    0      -5       0      -5
    0       0      -3      -9
>> ab(1,:) = ab(1,:) + 3/5*ab(2,:)
ab =

    1       0       0      -2
    0      -5       0      -5
    0       0      -3      -9
```

11.2.3 Reduced Row Echelon Form

The Gauss Jordan method results in a diagonal form; for example, for a 3×3 system:

$$\begin{bmatrix} a_{11} & a_{12} & a_{13} & b_1 \\ a_{21} & a_{22} & a_{23} & b_2 \\ a_{31} & a_{32} & a_{33} & b_3 \end{bmatrix} \rightarrow \begin{bmatrix} a'_{11} & 0 & 0 & b'_1 \\ 0 & a'_{22} & 0 & b'_2 \\ 0 & 0 & a'_{33} & b'_3 \end{bmatrix}$$

Reduced Row Echelon Form takes this one step further to result in all 1's rather than the a', so that the column of b's is the solution:

$$\begin{bmatrix} a_{11} & a_{12} & a_{13} & b_1 \\ a_{21} & a_{22} & a_{23} & b_2 \\ a_{31} & a_{32} & a_{33} & b_3 \end{bmatrix} \rightarrow \begin{bmatrix} 1 & 0 & 0 & b'_1 \\ 0 & 1 & 0 & b'_2 \\ 0 & 0 & 1 & b'_3 \end{bmatrix}$$

MATLAB has a built-in function to do this, called **rref**. For example, for the preceding example:

```
>> a = [1 3 0; 2 1 3; 4 2 3];
>> b = [1 6 3]';
>> ab = [a b];
>> rref(ab)
ans =

    1       0       0      -2
    0       1       0       1
    0       0       1       3
```

The solution is found from the last column, so $x_1 = -2$, $x_2 = 1$, and $x_3 = 3$. To get this in a column vector in MATLAB:

```
>> x = ans(:,end)
x =
    -2
     1
     3
```

11.2.4 Finding a Matrix Inverse by Reducing an Augmented Matrix

For a system of equations larger than a 2×2 system, one method of finding the inverse of a matrix A mathematically involves augmenting the matrix with an identity matrix of the same size, and then reducing it. The algorithm is:

- Augment the matrix with I: [A I]

- Reduce it to the form [I X]; X will be A^{-1}

For example, in MATLAB we can start with a matrix, augment it with an identity matrix, and then use the **rref** function to reduce it.

```
>> a = [1 3 0; 2 1 3; 4 2 3];
>> rref([a eye(size(a))])
ans =

    1.0000        0         0      -0.2000    -0.6000     0.6000
         0   1.0000         0       0.4000     0.2000    -0.2000
         0        0    1.0000            0     0.6667    -0.3333
```

In MATLAB, the **inv** function can be used to verify the result.

```
>> inv(a)
ans =
    -0.2000    -0.6000     0.6000
     0.4000     0.2000    -0.2000
          0     0.6667    -0.3333
```

11.3 SYMBOLIC MATHEMATICS

Symbolic mathematics means doing mathematics on symbols (not numbers!). For example, a + a is 2a. The symbolic math functions are in the Symbolic Math Toolbox in MATLAB. Toolboxes contain related functions and are add-ons to MATLAB. (Therefore, this may or may not be part of your own system.) The Symbolic Math Toolbox includes an alternative method for solving equations, and therefore is covered in this chapter.

To find out about the symbolic functions, **help** can be used:

```
help toolbox\symbolic
```

11.3.1 Symbolic Variables and Expressions

MATLAB has a type called **sym** for symbolic variables and expressions; these work with strings. For example, to create a symbolic variable a and perform the addition just described, first a symbolic variable would be created by passing the string 'a' to the **sym** function:

```
>> a = sym('a');
>> a+a
ans =
2*a
```

Symbolic variables can also store expressions. For example, the variables b and c store symbolic expressions:

```
>> b = sym('x^2');
>> c = sym('x^4');
```

All basic mathematical operations can be performed on symbolic variables and expressions (e.g., add, subtract, multiply, divide, raise to a power, etc.). Here are some examples:

```
>> c/b
ans =
x^2

>> b^3
ans =
x^6
>> c*b
ans =

x^6
>> b + sym('4*x^2')
ans =
5*x^2
```

It can be seen from the last example that MATLAB will collect like terms in these expressions, adding the x^2 and $4x^2$ to result in $5x^2$.

The following creates a symbolic expression by passing a string, but the terms are not automatically collected:

```
>> sym('z^3 + 2*z^3')
ans =
z^3 + 2*z^3
```

If, on the other hand, z is a symbolic variable to begin with, quotes are not needed around the expression, and the terms are automatically collected:

```
>> z = sym('z');
>> sym(z^3 + 2*z^3)
ans =
3*z^3
```

If you want to use multiple variables as symbolic variable names, the **syms** function is a shortcut instead of using **sym** repeatedly. For example,

```
>> syms x y z
```

does the same thing as

```
>> x = sym('x');
>> y = sym('y');
>> z = sym('z');
```

11.3.2 Simplification Functions

There are several functions that work with expressions, and simplify the terms. Not all expressions can be simplified, but the **simplify** function does whatever it can to simplify expressions, including gathering like terms. For example:

```
>> x = sym('x');
>> myexpr = cos(x)^2 + sin(x)^2
myexpr =
cos(x)^2+sin(x)^2
>> simplify(myexpr)
ans =
1
```

The functions **collect, expand**, and **factor** work with polynomial expressions. The **collect** function collects coefficients, for example,

```
>> x = sym('x');
>> collect(x^2 + 4*x^3 + 3*x^2)
ans =
4*x^2+4*x^3
```

The **expand** function will multiply out terms, and **factor** will do the reverse:

```
>> expand((x+2)*(x-1))
ans =
x^2+x-2

>> factor(ans)
ans =
(x+2)*(x-1)
```

The **subs** function will substitute a value for a symbolic variable in an expression. For example,

```
>> myexp = x^3 + 3*x^2 - 2
myexp =
x^3+3*x^2-2
>> x = 3;
>> subs(myexp,x)
ans =
    52
```

With symbolic math, MATLAB works by default with rational numbers, meaning that results are kept in fractional forms. For example, performing the addition 1/3 + 1/2 would normally result in a **double** value:

```
>> 1/3 + 1/2
ans =
    0.8333
```

However, by making the expression symbolic, the result is symbolic also. Any numeric function (e.g., **double**) could change that:

```
>> sym(1/3 + 1/2)
ans =
5/6

>> double(ans)
ans =
    0.8333
```

The **numden** function will return separately the numerator and denominator of a symbolic expression:

```
>> sym(1/3 + 1/2)
ans =
5/6

>> [n, d] = numden(ans)
n =
5
d =
6

>> [n, d] = numden((x^3 + x^2)/x)
n =
x^2*(x+1)
d =
x
```

11.3.3 Displaying Expressions

The **pretty** function will display these expressions using exponents; for example,

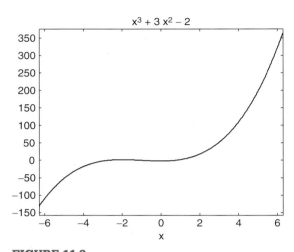

FIGURE 11.2

*Plot produced using **ezplot.***

```
>> b = sym('x^2')
b =
x^2
>> pretty(b)
            2
          x
```

There are several plot functions in MATLAB with names beginning with "ez" that perform the necessary conversions from symbolic expressions to numbers and plot them. For example, the function **ezplot** will draw a 2-D plot in the x-range from -2π to 2π, with the expression as the title (in pretty form). The expression

```
>> ezplot('x^3 + 3*x^2 - 2')
```

produces the figure shown in Figure 11.2.

The domain for the **ezplot** function can also be specified; for example, to change the x-axis to be from 0 to π, it is specified as a vector. The result is shown in Figure 11.3.

```
>> ezplot('cos(x)',[0 pi])
```

11.3.4 Solving Equations

We've seen already several methods for solving simultaneous linear equations, using a matrix representation. MATLAB can also solve sets of equations using symbolic math.

FIGURE 11.3

*Result from **ezplot** with custom x-axis.*

The function **solve** solves an equation and returns the solution(s) as symbolic expressions. The solution can be converted to numbers using any numeric function, such as **double**:

```
>> x = sym('x');
>> solve('2*x^2 + x = 6')
ans =
3/2
 -2

>> double(ans)
ans =
  1.5000
 -2.0000
```

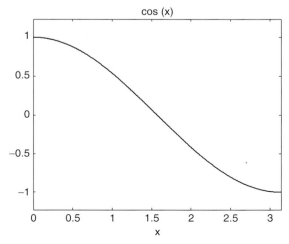

If an expression is passed to the **solve** function rather than an equation, it will set the expression equal to 0 and solve the resulting equation. For example, this will solve $3x^2 + x = 0$:

```
>> solve('3*x^2 + x')
ans =
     0
  -1/3
```

If there is more than one variable, MATLAB chooses which to solve for. In the following example, the equation $ax^2 + bx = 0$ is solved. There are three variables. As can be seen from the result, which is given in terms of a and b, the equation was solved for x. MATLAB has rules built in that specify how to choose which variable to solve for. For example, x will always be the first choice if it is in the equation or expression.

```
>> solve('a*x^2 + b*x')
ans =
     0
  -b/a
```

However, it is possible to specify which variable to solve for:

```
>> solve('a*x^2 + b*x', 'b')
ans =
  -a*x
```

MATLAB can also solve sets of equations. In this example, the solutions for x, y, and z are returned as a structure consisting of fields for x, y, and z. The individual solutions are symbolic expressions stored in fields of the structure.

```
>> solve('4*x-2*y+z=7', 'x+y+5*z=10', '-2*x+3*y-z=2')
ans =
    x: [1x1 sym]
    y: [1x1 sym]
    z: [1x1 sym]
```

To refer to the individual solutions, which are in the structure fields, the dot operator is used.

```
>> x = ans.x
x =
124/41
```

```
>> y = ans.y
y =
121/41
```

```
>> z = ans.z
z =
33/41
```

The **double** function can then be used to covert the symbolic expressions to numbers, and store the results from the three unknowns in a vector.

```
>> double([x y z])
ans =
    3.0244        2.9512        0.8049
```

SUMMARY

Common Pitfalls

- Confusing matrix multiplication and array multiplication. Array operations, including multiplication, division, and exponentiation, are performed term-by-term (so the arrays must have the same size); the operators are .*, ./, .\, and .^. For matrix multiplication to be possible, the inner dimensions must agree and the operator is *.
- Forgetting that to augment one matrix with another, the number of rows must be the same in each (but not necessarily the number of columns).

Programming Style Guidelines

- When working with symbolic expressions, it is generally easier to make all variables symbolic variables to begin with.

MATLAB Functions and Commands			
trace	dot	sym	subs
diag	cross	syms	numden
eye	ones	simplify	pretty
triu	magic	collect	ezplot
tril	det	expand	solve
inv	rref	factor	

MATLAB Operators
* matrix multiplication
\ matrix division

Exercises

1. For the following matrices A, B, and C:

$$
A = \begin{bmatrix} 1 & 4 \\ 3 & 2 \end{bmatrix} \quad
B = \begin{bmatrix} 2 & 1 & 3 \\ 1 & 5 & 6 \\ 3 & 6 & 0 \end{bmatrix} \quad
C = \begin{bmatrix} 3 & 2 & 5 \\ 4 & 1 & 2 \end{bmatrix}
$$

- Which are symmetric?
- For all square matrices, give their trace.

- Give the result of 3*A.
- Give the result of A*C.
- Are there any other matrix multiplications that can be performed? If so, list them.

2. Given the following matrices:

$$
\begin{array}{ccc}
A & B & C
\end{array}
$$

$$
\begin{bmatrix} 3 & 2 & 1 \\ 0 & 5 & 2 \\ 1 & 0 & 3 \end{bmatrix} \quad \begin{bmatrix} 2 \\ 1 \\ 3 \end{bmatrix} \quad \begin{bmatrix} 1 & 0 & 0 \\ 0 & 1 & 0 \\ 0 & 0 & 1 \end{bmatrix}
$$

Perform the following MATLAB operations, if they can be done. If not, explain why.

```
A * B

B * A

I + A

A .* I

trace(A)
```

3. Write a function *issquare* that will receive an array argument, and will return 1 for true if it is a square matrix, or 0 for false if it is not.

4. Write a function *mydiag* that will receive an array argument, and will return a vector consisting of the main diagonal (without using the built-in **diag** function). Note: this is possible only if the argument is a square matrix, so the function should first check this by calling the *issquare* function from above. If the argument is a square matrix, the function will return the diagonal; otherwise, it will return an empty vector.

5. Write a function that will receive a square matrix as an input argument, and will return a row vector containing the diagonal of the matrix. If the function is called with a vector of two variables on the left-hand side of the assignment, the function will also return the trace of the matrix. (Note: It will return the trace *only* if there are two variables on the left-hand side of the assignment.) You may assume that the matrix is square. The function *must* preallocate the diagonal vector to the correct size.

6. Write a function *randdiag* that will return an n × n diagonal matrix, with random integers each in the range from low to high on the diagonal. Three arguments are passed to the function: the value of n, low, and high, in that order.

7. Write a function *myeye* to return an n × n identity matrix.

8. Write a function *myupp* that will receive an integer argument n, and will return an n × n upper triangular matrix of random integers.

9. Write a function to determine whether or not a square matrix is a diagonal matrix. This function will return 1 for true if it is, or 0 if not.

10. Write a function *mymatsub* that will receive two matrix arguments and will return the result of subtracting the elements in one matrix from another (by looping and subtracting term by term). If it is not possible to subtract, return an empty matrix.

11. Write a function to receive a matrix and return its transpose (for more programming practice, do not use the built-in operator for the transpose).

12. We have already seen the **zeros** function, which returns a matrix of all 0's. Similarly, there is a function **ones** that returns a matrix of all 1's. Note: No, there aren't functions called twos, threes, and such (just **ones** and **zeros**!). However, write a *fives* function that will receive two arguments for the number of rows and columns and will return a matrix with that size of all 5's.

13. The function **magic(n)** returns an $n \times n$ magic matrix, which is a matrix for which the sum of all rows, columns, and the diagonal are the same. Investigate this built-in function.

14. The function **pascal(n)** returns an $n \times n$ matrix made from Pascal's triangle. Investigate this built-in function, and then write your own.

15. Rewrite the following system of equations in matrix form:

$$4x_1 - x_2 + 3x_4 = 10$$
$$-2x_1 + 3x_2 + x_3 - 5x_4 = -3$$
$$x_1 + x_2 - x_3 + 2x_4 = 2$$
$$3x_1 + 2x_2 - 4x_3 = 4$$

Set it up in MATLAB and use any method to solve.

16. For the following 2×2 system of equations:

$$-3x_1 + x_2 = -4$$
$$-6x_1 + 2x_2 = 4$$

- In MATLAB, rewrite the equations as equations of straight lines and plot them to find the intersection.
- Solve for one of the unknowns and then substitute into the other equation to solve for the other unknown.
- Find the determinant D.
- How many solutions are there? One? None? Infinite?

17. For the following 2×2 system of equations:

$$-3x_1 + x_2 = 2$$
$$-6x_1 + 2x_2 = 4$$

- Rewrite the equations as equations of straight lines and plot them to find the intersection.
- Solve for one of the unknowns and then substitute into the other equation to solve for the other unknown.
- Find the determinant D.
- How many solutions are there? One? None? Infinite?

18. Write a function to return the determinant of a 2×2 matrix.
19. Write a function to return the inverse of a 2×2 matrix.
20. Given the following 2×2 system of equations:

$$3x_1 + x_2 = 2$$
$$2x_1 = 4$$

Use all methods presented in the text to solve it, and to visualize the solution. Do all the math by hand, and then also use MATLAB.

21. ERO Practice: Show the result of each of the following EROs:

$$\begin{bmatrix} 4 & 2 & 3 \\ 1 & 4 & 0 \\ 2 & 5 & 3 \end{bmatrix} \frac{1}{4}r1 \rightarrow r1$$

$$\begin{bmatrix} 4 & 2 & 3 \\ 1 & 4 & 0 \\ 2 & 5 & 3 \end{bmatrix} r2 \longleftrightarrow r3$$

$$\begin{bmatrix} 4 & 2 & 3 \\ 1 & 4 & 0 \\ 2 & 5 & 3 \end{bmatrix} r3 - 2r2 \rightarrow r3$$

22. For the following 2×2 system of equations:

$$3x_1 + 2x_2 = 4$$
$$x_1 = 2$$

- Write this in matrix form.
- Using the method for 2×2 systems, find the determinant D.
- Use D to find the inverse of A.
- Use the Gauss elimination method to find the solution.
- Use the Gauss-Jordan method to solve.
- Check your work in MATLAB.

23. For the following set of equations:

$$2x_1 + 2x_2 + x_3 = 2$$
$$x_2 + 2x_3 = 1$$
$$x_1 + x_2 + 3x_3 = 3$$

- Put this in the augmented matrix [A|b].
- Solve using Gauss.
- Solve using Gauss-Jordan.
- In MATLAB, create the matrix A and vector b. Find the inverse and determinant of A. Solve for x.

24. Given the following system of equations:

$$\begin{aligned} x_1 - 2x_2 + x_3 &= 2 \\ 2x_1 - 5x_2 + 3x_3 &= 6 \\ x_1 + 2x_2 + 2x_3 &= 4 \\ 2x_1 + 3x_3 &= 6 \end{aligned}$$

Write this in matrix form and use either Gauss or Gauss-Jordan to solve it. Check your answer using MATLAB.

25. Write a function that will augment a matrix with an identity matrix of the appropriate dimensions, without using any built-in functions (except **size**). This function will receive a matrix argument, and will return the augmented matrix.

26. Write a function *myrrefinv* that will receive a square matrix A as an argument, and will return the inverse of A. The function cannot use the built-in inv function; instead, it must augment the matrix with I and use **rref** to reduce it to the form [I A^{-1}]. Here are examples of calling it:

```
>> a =[4 3 2; 1 5 3; 1 2 3]
a =

     4          3          2
     1          5          3
     1          2          3
>> inv(a)
ans =

     0.3000     -0.1667     -0.0333
     0           0.3333     -0.3333
    -0.1000     -0.1667      0.5667

>> myrrefinv(a)
ans =

     0.3000     -0.1667     -0.0333
     0           0.3333     -0.3333
    -0.1000     -0.1667      0.5667

>> disp(myrrefinv(a))

     0.3000     -0.1667     -0.0333
     0           0.3333     -0.3333
    -0.1000     -0.1667      0.5667
```

27. Solve the simultaneous equations $x - y = 2$ and $x^2 + y = 0$ using **solve**. Plot the corresponding functions, $y = x - 2$ and $y = -x^2$, on the same graph with an x range from −5 to 5.

28. For the following set of equations:

$$\begin{aligned} 2x_1 + 2x_2 + x_3 &= 2 \\ x_2 + 2x_3 &= 1 \\ x_1 + x_2 + 3x_3 &= 3 \end{aligned}$$

- In MATLAB, create the coefficient matrix A and vector b. Solve for x using the inverse, using the built-in function.
- Create the augmented matrix [A|b] and solve using the **rref** function.

Write this in symbolic form and solve using the **solve** function. From the symbolic solution, create a vector of the numerical (**double**) equivalents.

29. For the following system of equations:

$$4x_1 - x_2 + 3x_4 = 10$$
$$-2x_1 + 3x_2 + x_3 - 5x_4 = -3$$
$$x_1 + x_2 - x_3 + 2x_4 = 2$$
$$3x_1 + 2x_2 - 4x_3 = 4$$

Use the **solve** function to solve it. Verify the answer using any other method (in MATLAB!).

30. Biomedical engineers are developing an insulin pump for diabetics. To do this, it is important to understand how insulin is cleared from the body after a meal. The concentration of insulin at any time t is described by the equation

$$C = C_0 e^{-30t/m}$$

where C_0 is the initial concentration of insulin, t is the time in minutes, and m is the mass of the person in kg. Use **solve** to determine for a person whose mass is 65 kg how long it will take an initial concentration of 90 to reduce to 10. Use **double** to get your result in minutes.

31. To analyze electric circuits, it is often necessary to solve simultaneous equations. To find the voltages Va, Vb, and Vc at nodes a, b, and c, the equations are:

```
2(Va-Vb) + 5(Va-Vc) - e-t = 0
2(Vb - Va) + 2Vb + 3(Vb - Vc) = 0
Vc = 2 sin(t)
```

Find out how to use the **solve** function to solve for Va, Vb, and Vc so that the solution will be returned in terms of t.

32. The reproduction of cells in a bacterial colony is important for many environmental engineering applications such as wastewater treatments. The formula

$$\log(N) = \log(N_0) + t/T \log(2)$$

can be used to simulate this, where N_0 is the original population, N is the population at time t, and T is the time it takes for the population to double. Use the **solve** function to determine the following: if $N_0 = 10^2$, $N = 10^8$, and t = 8 hours, what will be the doubling time T? Use **double** to get your result in hours.

33. We have seen that a square matrix is symmetric if $a_{ij} = a_{ji}$ for all i, j. We say that a square matrix is **skew symmetric** if $a_{ij} = -a_{ji}$ for all i, j. Notice that this means that all the values on the diagonal must be 0. Write a function that will receive a square matrix as an input argument, and will return 1 for logical true if the matrix is skew symmetric or 0 for false if not.

Basic Statistics, Searching, and Sorting

CONTENTS

There are a lot of statistical analyses that can be performed on data sets. In the MATLAB® software, the statistical functions are in the data analysis help topic called **datafun**.

In general, we will write a data set of *n* values as

$$x = \{x_1, x_2, x_3, x_4, \ldots, x_n\}$$

In MATLAB, this will generally be represented as a row vector called x.

Statistics can be used to characterize properties of a data set. For example, consider a set of exam grades {33, 75, 77, 82, 83, 85, 85, 91, 100}. What is a "normal", or "expected", or "average" exam grade? There are several ways that this could be interpreted. Perhaps the most common is the *mean* grade, which is found by summing the grades and dividing by the number of them (the result of that would be 79). Another way of interpreting that would be the grade found the most often, which would be 85. Also, the value in the middle of the list, 83, could be used. Another property that is useful to know is how spread out the data values are within the data set.

This section will cover some simple statistics, as well as set operations that can be performed on data sets. Some statistical functions require that the data set be *sorted*, so sorting will also be covered. Using *index vectors* is a way of representing the data in order, without physically sorting the data set. Finally, *searching* for values within a data set or a database is useful, so some basic searching techniques will be explained.

12.1 STATISTICAL FUNCTIONS

MATLAB has built-in functions for many statistics; the simplest of which we have already seen; for example, **min** and **max** to find the minimum or maximum value in a data set.

```
>> x = [9 10 10 9 8 7 3 10 9 8 5 10];
>> min(x)
ans =
     3
>> max(x)
ans =
     10
```

Both of these functions also return the index of the smallest or largest value; if there is more than one occurrence, it returns the first. For example, in the following data set, 10 is the largest value; it is found in three elements in the vector but the index returned is the first element in which it is found (which is 2):

```
>> x = [9 10 10 9 8 7 3 10 9 8 5 10];
>> [maxval, maxind] = max(x)
maxval =
     10
maxind =
     2
```

For matrices, the **min** and **max** functions operate columnwise by default:

```
>> mat = randint(2,4,[1 20])
mat =
     9        10       17        5
    19         9       11       14
>> min(mat)
ans=
```

```
    9       9       11       5
>> [minval, minind] = min(mat)
minval =
    9       9       11       5
minind =
    1       2        2       1
```

To find the minimum (or maximum) for each row, the dimension of 2 (which is how MATLAB refers to rows) can be specified as the third argument to the **min** (or **max**) function; the second argument *must* be an empty vector:

```
>> min(mat,[],2)
ans =
    5
    9
```

These functions can also compare vectors or matrices and return the minimum (or maximum) values from corresponding elements. For example, the following iterates through all elements in the two vectors, comparing corresponding elements and returning the minimum for each:

```
>> x = [3 5 8 2 11];
>> y = [2 6 4 5 10];
>> min(x,y)
ans =
    2       5       4       2       10
```

Some of the other functions in the **datafun** help topic that have been described already include **sum**, **prod**, **cumsum**, **cumprod**, and **hist**. Other statistical operations, and the functions that perform them in MATLAB, will be described in the rest of this section.

12.1.1 Mean

The *arithmetic mean* of a data set is what is usually called the *average* of the values; in other words, the sum of the values divided by the number of values in the data set. Mathematically, we would write this as

$$\frac{\sum_{i=1}^{n} x_i}{n}.$$

The Programming Concept

Calculating a mean, or average, normally would be interpreted as looping through the elements of a vector, adding them together, and then dividing by the number of elements:

mymean.m

```
function outv = mymean(vec)
% Calculates and returns the mean of a vector
mysum = 0;
for i=1:length(vec)
    mysum = mysum + vec(i);
end
outv = mysum/length(vec);
```

```
>> x = [9 10 10 9 8 7 3 10 9 8 5 10];
>> mymean(x)
ans =
    8.1667
```

The Efficient Method

There is a built-in function, **mean**, in MATLAB to accomplish this:

```
>> mean(x)
ans =
    8.1667
```

For a matrix, the **mean** function operates columnwise. To find the mean of each row, the dimension of 2 is passed as the second argument to the function, as is the case with the functions **sum, prod, cumsum,** and **cumprod** (the [] as a middle argument is not necessary for these functions like it is for **min** and **max**).

```
>> mat = randint(3,3,[1 10])
mat =
     8     9     3
    10     2     3
     6    10     9
>> mean(mat)
ans =

     8     7     5
>> mean(mat,2)
ans =
    6.6667
    5.0000
    8.3333
```

Sometimes a value that is much larger or smaller than the rest of the data (called an *outlier*) can throw off the mean; for example, in the following all the numbers in the data set are in the range from 3 to 10, with the exception of the 100 in the middle. Because of this outlier, the mean of the values in this vector is actually larger than any of the other values in the vector.

```
>> xwithbig = [9 10 10 9 8 100 7 3 10 9 8 5 10];
>> mean(xwithbig)
ans =
     15.2308
```

In order to handle this, sometimes the minimum and maximum values from a data set are discarded before the mean is computed. In this example, a logical vector indicating which elements are neither the largest nor smallest value is used to index into the original data set, resulting in removing the minimum and the maximum.

```
>> xwithbig = [9 10 10 9 8 100 7 3 10 9 8 5 10];
>> length(xwithbig)
ans =
     13
>> newx = xwithbig(xwithbig = min(xwithbig) & ...
xwithbig = max(xwithbig))
newx =
     9    10    10    9    8    7    10    9    8    5    10
>> length(newx)
ans =
     11
```

Instead of just removing the minimum and maximum values, sometimes the largest and smallest 1% or 2% of values are removed, especially if the data set is very large.

There are several other means that can be computed. The *harmonic mean* of the *n* values in a vector or data set x is defined as

$$\frac{n}{\dfrac{1}{x_1} + \dfrac{1}{x_2} + \dfrac{1}{x_3} + \ldots + \dfrac{1}{x_n}}$$

This could be implemented in an anonymous function using the built-in **sum** function. For example, the following anonymous function calculates this, and stores the handle in a variable called *harmhand*.

```
>> harmhand = @ (x) length(x) / sum ( 1 ./ x );
>> x = [9 10 10 9 8 7 3 10 9 8 5 10];
>> harmhand(x)
ans =
     7.2310
```

The *geometric mean* of the *n* values in a vector x is defined as the nth root of the product of the data set values.

$$\sqrt[n]{x_1 * x_2 * x_3 \ldots * x_n}$$

The following anonymous function implements this definition, using **prod**:

```
>> geomhand = @ (x) prod(x)^(1/length(x));
>> geomhand(x)
ans =
    7.7775
```

Note: Statistics Toolbox™ has functions for these means, called **harmmean** and **geomean**, as well as a function **trimmean**, which trims the highest and lowest 2% of data values.

12.1.2 Variance and Standard Deviation

The *standard deviation* and *variance* are ways of determining the spread of the data. The variance is usually defined in terms of the arithmetic mean as:

$$var = \frac{\sum_{i=1}^{n}(x_i - mean)^2}{n-1}$$

Sometimes, however, the denominator is defined as *n* rather than *n* − 1. The default definition used by MATLAB is given by the previous equation, so we will use that definition here.

For example, for the vector [4, 6, 1, 5], there are *n* = 4 values so *n* − 1 is 3. Also, the mean of this data set is 4. The variance would be

$$var = \frac{(4-4)^2 + (6-4)^2 + (1-4)^2 + (5-4)^2}{3} = \frac{0+4+9+1}{3} = 4.67$$

The built-in function to calculate the variance is called **var**:

```
>> shortx = [4 6 1 5];
>> myvar = var(shortx)
myvar =
    4.6667
```

The standard deviation is the square root of the variance:

$$sd = \sqrt{var}$$

The built-in function in MATLAB for the standard deviation is called **std**; the standard deviation can be found either as the **sqrt** of the variance, or using **std**:

```
>> shortx = [2 5 1 4];
>> myvar = var(shortx)
myvar =
    3.3333
>> sqrt(myvar)
ans =
    1.8257
>> std(shortx)
ans =
    1.8257
```

The less spread out the numbers are, the smaller the standard deviation will be, since it is a way of determining the spread of the data. Likewise, the more spread out the numbers are, the larger the standard deviation will be. For example here are two data sets that have the same number of values and also the same mean, but the standard deviations are quite different:

```
>> x1 = [9 10 9.4 9.6];
>> mean(x1)
ans =
    9.5000
>> std(x1)
ans =
    0.4163
>> x2 = [2 17 -1.5 20.5];
>> mean(x2)
ans =
    9.5000
>> std(x2)
ans =
    10.8704
```

12.1.3 Mode

The *mode* of a data set is the value that appears most frequently. The built-in function in MATLAB for this is called **mode**.

```
>> x = [9 10 10 9 8 7 3 10 9 8 5 10];
>> mode(x)
ans =
    10
```

If there is more than one value with the same (highest) frequency, the smaller value is the mode. In the following case, since 3 and 8 appear twice in the vector, the smaller value (3) is the mode:

```
>> x = [3 8 5 3 4 1 8];
>> mode(x)
ans =
       3
```

If no value appears more frequently than any other, the smallest value in the vector will be the mode of the vector.

```
>> shortx = [2 5 1 4];
>> mode(shortx)
ans =
       1
```

12.1.4 Median

The *median* is defined only for a data set that has been *sorted* first, meaning that the values are in order. The median of a sorted set of n data values is defined as the value in the middle, if n is odd, or the average of the two values in the middle, if n is even. For example, for the vector [1 4 5 9 12], the middle value is 5. The function in MATLAB is called **median**:

```
>> median([1 4 5 9 12 ])
ans =
       5
```

For the vector [1 4 5 9 12 33], the median is the average of the 5 and 9 in the middle:

```
>> median([1 4 5 9 12 33])
ans =
       7
```

 PRACTICE 12.1

For the vector [1 1 3 6 9], find the

- minimum
- maximum
- arithmetic mean
- geometric mean
- harmonic mean
- variance
- standard deviation
- mode
- median

If the vector is not in sorted order to begin with, the **median** function will still return the correct result (it will sort the vector automatically). For example, scrambling the order of the values in the first example will still result in a median value of 5.

```
>> median([9 4 1 5 12])
ans =
       5
```

12.2 SET OPERATIONS

MATLAB has several built-in functions that perform *set operations* on vectors. These include **union**, **intersect**, **unique**, **setdiff**, and **setxor**. All these functions can be useful when working with data sets. Additionally, there are two **is** functions that work on sets, **ismember** and **issorted**.

For example, given the following vectors:

```
>> v1 = 2:6
v1 =
     2     3     4     5     6
>> v2 = 1:2:7
v2 =
     1     3     5     7
```

The **union** function returns a vector that contains all the values from the two input argument vectors, without repeating any.

```
>> union(v1,v2)
ans =
     1     2     3     4     5     6     7
```

The **intersect** function instead returns all the values that can be found in *both* of the input argument vectors.

```
>> intersect(v1,v2)
ans =
     3     5
```

The **setdiff** function receives two vectors as input arguments, and returns a vector consisting of all the values that are contained in the first vector argument but not the second. Therefore, the order of the two input arguments is important.

```
>> setdiff(v1,v2)
ans =
     2     4     6
>> setdiff(v2,v1)
ans =
     1     7
```

The function **setxor** receives two vectors as input arguments, and returns a vector consisting of all the values from the two vectors that are not in the intersection of these two vectors. In other words, it is the union of the two vectors obtained using **setdiff** earlier!

```
>> setxor(v1,v2)
ans =
     1     2     4     6     7
>> union(setdiff(v1,v2), setdiff(v2,v1))
ans =
     1     2     4     6     7
```

The set function **unique** returns all the unique values from a set argument:

```
>> v3 = [1:5 3:6]
v3 =
    1    2    3    4    5    3    4    5    6
>> unique(v3)
ans =
    1    2    3    4    5    6
```

Many of the set functions return vectors that can be used to index into the original vectors as optional output arguments; for example, for the two vectors *v1* and *v2* defined previously as:

```
>> v1
v1 =
    2    3    4    5    6
>> v2
v2 =
    1    3    5    7
```

The **intersect** function returns, in addition to the vector containing the values in the intersection of *v1* and *v2*, an index vector into *v1* and an index vector into *v2* such that *outvec* is the same as *v1(index1)* and also *v2(index2)*.

```
>> [outvec, index1, index2] = intersect(v1,v2)
outvec =
    3    5

index1 =
    2    4

index2 =
    2    3
```

Using these vectors to index into *v1* and *v2* will return the values from the intersection. For example, this expression returns the second and fourth elements of *v1*:

```
>> v1(index1)
ans =
    3    5
```

This returns the second and third elements of *v2*:

```
>> v2(index2)
ans =
    3    5
```

The function **ismember** receives two vectors as input arguments, and returns a logical vector that is the same length as the first argument, containing 1 for true if the element in the first vector is also in the second, or 0 for false if not. The order of the arguments matters for this function.

```
>> v1
v1 =
     2      3      4      5      6
>> v2
v2 =
     1      3      5      7
>> ismember(v1,v2)
ans =
     0      1      0      1      0
>> ismember(v2,v1)
ans =
     0      1      1      0
```

Using the result from the **ismember** function as an index into the first vector argument will return the same result as the **intersect** function.

```
>> logv = ismember(v1,v2)
logv =
     0      1      0      1      0
>> v1(logv)
ans =
     3      5
>> logv = ismember(v2,v1)
logv =
     0      1      1      0
>> v2(logv)
ans =
     3      5
```

The **issorted** function will return 1 for logical true if the argument is sorted in *ascending* order (lowest to highest), or 0 for false if not.

```
>> v3 = [1:5 3:6]
v3 =
     1      2      3      4      5      3      4      5      6
>> issorted(v3)
ans =
     0
```

```
>> issorted(v1)
ans =
     1
>> vec = 4:-1:1
vec =
     4        3        2        1
>> issorted(vec)
ans =
     0
```

In the next section, we will see how to sort a vector.

 PRACTICE 12.2

Create two vector variables *vec1* and *vec2* that contain seven random integers, each in the range from 1 to 20. Do each of the following operations by hand first, and then check in MATLAB:

- union
- intersection
- setdiff
- setxor
- unique (for each)

12.3 SORTING

Sorting is the process of putting a list in order; either *descending* (highest to lowest), or *ascending* (lowest to highest) order. For example, here is a list of *n* integers, visualized as a column vector.

1	85
2	70
3	100
4	95
5	80
6	91

We want to sort this in ascending order in place, in other words, rearranging this vector, not creating another. One basic algorithm would be:

- Look through the vector to find the smallest number, and then put it in the first element in the vector. How? By exchanging it with the number currently in the first element.

- Then, scan the rest of the vector (from the second element down) looking for the next smallest (or, the smallest in the rest of the vector). When found, put it in the first element of the rest of the vector.

- Continue doing this for the rest of the vector. Once the next-to-last number has been placed in the correct location in the vector, by default the last number has been, also.

This table shows the progression. The left-hand column shows the original vector. The second column (from the left) shows that the smallest number, the 70, is now in the first element in the vector. It was put there by exchanging with what had been in the first element, 85. This continues element-by-element, until the vector has been sorted.

85	70	70	70	70	70
70	85	80	80	80	80
100	100	100	85	85	85
95	95	95	95	91	91
80	80	85	100	100	95
91	91	91	91	95	100

This is called the *selection sort*; it is just one of many different sorting algorithms.

The Programming Concept

The following function implements the selection sort to sort a vector:

mysort.m

```
function outv = mysort(vec)
% This function sorts a vector using the selection sort
% Loop through the elements in the vector to end-1
for i = 1:length(vec)-1
    low = i;
    %Select the lowest number in the rest of the vector
    for j=i+1:length(vec)
      if vec(j) < vec(low)
          low = j;
      end
    end
    % Exchange elements
    temp = vec(i);
    vec(i) = vec(low);
    vec(low) = temp;
end
outv = vec;
```

```
>> vec = [85 70 100 95 80 91];
>> vec = mysort(vec)
vec =
    70      80      85      91      95      100
```

The Efficient Method

MATLAB has a built-in function, **sort**, that will sort a vector in ascending order:

```
>> vec = [85 70 100 95 80 91];
>> vec = sort(vec)
vec =
    70      80      85      91      95     100
```

Descending order can also be specified; for example:

```
>> sort(vec,'descend')
ans =
   100      95      91      85      80      70
```

For matrices, the **sort** function will by default sort each column. To sort by rows, the dimension 2 is specified. For example,

```
>> mat
mat =
    4      6      2
    8      3      7
    9      7      1
>> sort(mat) % sorts by column
ans =
    4      3      1
    8      6      2
    9      7      7
>> sort(mat,2) % sorts by row
ans =
    2      4      6
    3      7      8
    1      7      9
```

12.3.1 Sorting Vectors of Structures

When working with a vector of structures, it is common to sort based on a particular field within the structures. For example, recall the vector of structures used to store information on different software packages that was created in Chapter 7.

packages			
item_no	cost	price	code
1	123	19.99	39.95	'g'
2	456	5.99	49.99	'l'
3	587	11.11	33.33	'w'

Here is a function that sorts this vector of structures in ascending order based on the *price* field.

mystructsort.m

```
function outv = mystructsort(structarr)
%This function sorts the packages struct
% based on the price field
for i = 1:length(structarr)-1
    low = i;
    for j=i+1:length(structarr)
      if structarr(j).price < structarr(low).price
        low = j;
      end
    end
    % Exchange elements
    temp = structarr(i);
    structarr(i) = structarr(low);
    structarr(low) = temp;
end
outv = structarr;
```

Notice that only the *price* field is compared in the sort algorithm, but the entire structure is exchanged. That is so that each element in the vector, which is a structure of information about a particular software package, remains intact.

Recall that we created a function *printpackages* also in Chapter 7, which prints the information in a nice table format. Calling the *mystructsort* function and also the function to print will demonstrate this:

```
>> printpackages(packages)
Item #      Cost      Price     Code
  123       19.99     39.95      g
  456        5.99     49.99      1
  587       11.11     33.33      w
>> pack_by_price = mystructsort(packages);
>> printpackages(pack_by_price)
Item #      Cost      Price     Code
  587       11.11     33.33      w
  123       19.99     39.95      g
  456        5.99     49.99      1
```

This function sorts the structures based only on the *price* field. A more general function is shown next, which receives a string that is the name of the field. The function checks first to make sure that the string that is passed is a valid field name for the structure. If it is, it sorts based on that field, and if not it prints an error message and returns an empty vector. Strings are created consisting of the name of the vector variable, followed by parentheses containing the element number, the period, and finally the name of the field. The strings are created using square brackets to concatenate the pieces of the string, and the **int2str** function is used to convert the element number to a string. Then, using the **eval** function, the vector elements are compared to determine the lowest.

general_pack_sort.m

```
function outv = general_pack_sort(inputarg, fname)
%This function sorts a vector of package structs
% based on the field name passed as an input argument
if isfield(inputarg,fname)
   for i = 1:length(inputarg)-1
      low = i;
      for j=i+1:length(inputarg)
        if eval(['inputarg(' int2str(j) ').' fname]) < ...
          eval(['inputarg(' int2str(low) ').' fname])
          low = j;
        end
      end
      % Exchange elements
      temp = inputarg(i);
      inputarg(i) = inputarg(low);
      inputarg(low) = temp;
   end
  outv = inputarg;
else
  disp('Sorry, not a correct field name')
  outv = [];
end
```

Here are examples of calling the function:

```
>> pack_by_price = general_pack_sort(packages,'price');
>> printpackages(pack_by_price)
```

```
Item #       Cost      Price     Code
  587       11.11      33.33      w
  123       19.99      39.95      g
  456        5.99      49.99      1
>> pack_by_cost = general_pack_sort(packages, 'cost');
>> printpackages(pack_by_cost)
Item #       Cost      Price     Code
  456        5.99      49.99      1
  587       11.11      33.33      w
  123       19.99      39.95      g
>> pack_by_profit = general_pack_sort(packages, 'profit')
Sorry, not a correct field name
pack_by_profit =
[]
```

QUICK QUESTION!

Is this sort function general? Would it work for any vector of structures, not just one configured like *packages*?

Answer: It is fairly general. It will work for any vector of structures. However, the comparison will work for only numerical or character fields. So, as long as the field is a number or character, this function will work for any vector of structures. If the field is a vector itself (including a string), it will not work.

12.3.2 Sorting Strings

For a matrix of strings, the **sort** function works exactly as shown previously for numbers. For example,

```
>> words = char('Hello', 'Howdy', 'Hi', 'Goodbye', 'Ciao')
words =
Hello
Howdy
Hi
Goodbye
Ciao
```

The following sorts column-by-column using the ASCII equivalents of the characters. It can be seen from the results that the space character comes before the letters of the alphabet in the character encoding:

```
>> sort(words)
ans =
Ce
Giad
Hildb
```

```
Hoolo
Howoyye
```

To sort on the rows instead the second dimension must be specified.

```
>> sort(words,2)
ans =
   Hello
   Hdowy
       Hi
Gbdeooy
   Caio
```

It can be seen by this that the blank space comes before the letters of the alphabet in the character encoding, and also that the uppercase letters come before the lowercase letters.

How could the strings be sorted alphabetically? MATLAB has a function **sortrows** that will do this. The way it works is that it examines the strings column-by-column starting from the left. If it can determine which letter comes first, it picks up the entire string and puts it in the first row. In this example, the first two strings are placed based on the first character, C and G. For the other three strings, they all begin with H so the next column is examined. In this case the strings are placed based on the second character, e, i, o.

```
>> sortrows(words)
ans =
Ciao
Goodbye
Hello
Hi
Howdy
```

The **sortrows** function sorts each row as a block, or group, and it will also work on numbers. In this example the rows beginning with 3 and 4 are placed first. Then, for the rows beginning with 5, the values in the second column (6 and 7) determine the order.

```
>> mat = [5 7 2; 4 6 7; 3 4 1; 5 6 2]
mat =

     5     7     2
     4     6     7
     3     4     1
     5     6     2
>> sortrows(mat)
ans =
```

```
3    4    1
4    6    7
5    6    2
5    7    2
```

To sort a cell array of strings, the **sort** function can be used. For example,

```
>> engcellnames = {'Chemical','Mechanical',...
      'Biomedical','Electrical', 'Industrial'};
>> sort(engcellnames')
ans =
    'Biomedical'
    'Chemical'
    'Electrical'
    'Industrial'
    'Mechanical'
```

12.4 INDEXING

Indexing is an alternative to sorting a vector. With indexing, the vector is left in its original order. An index vector is used to point to the values in the original vector in the desired order.

For example, for a vector of exam grades:

grades					
1	2	3	4	5	6
85	70	100	95	80	91

In ascending order, the lowest grade is in element 2, the next lowest is in element 5, and so on. The index vector *grade_index* gives this order:

grade_index					
1	2	3	4	5	6
2	5	1	6	4	3

The index vector is then used as the indices for the original vector. To get the *grades* vector in ascending order, the indices used would be grades(2), grades(5), and so on. Using the index vector to accomplish this, grades(grade_index(1)) would be the lowest grade, 70, and grades(grade_index(2)) would be the second lowest grade. In general, grades(grade_index(i)) would be the ith lowest grade.

To create these in MATLAB:

```
>> grades = [85 70 100 95 80 91];
>> grade_index = [2 5 1 6 4 3];
>> grades(grade_index)
ans =
    70    80    85    91    95   100
```

However, instead of creating the index vector manually as shown here, the procedure to initialize the index vector is to use a sort function. The algorithm is:

- Initialize the values in the index vector to be the indices 1,2, 3, ...

- Use any sort algorithm, but compare the elements in the original vector using the index vector to index into it (e.g., using `grades(grade_index(i))` as shown earlier).

- When the sort algorithm calls for exchanging values, exchange the elements in the index vector, not in the original vector.

Here is a function that implements this algorithm:

createind.m

```
function indvec = createind(vec)
%This function creates index vector in ascending order

% Initialize the index vector
len = length(vec);
indvec = 1:len;
for i = 1:len-1
    low = i;
    for j=i+1:len
        % Compare values in the original vector
        if vec(indvec(j)) < vec(indvec(low))
            low = j;
        end
    end
    % Exchange elements in the index vector
    temp = indvec(i);
    indvec(i) = indvec(low);
    indvec(low) = temp;
end
```

For example, for the grades vector just given:

```
>> clear grade_index
>> grade_index = createind(grades)
grade_index =
     2     5     1     6     4     3
```

```
>> grades(grade_index)
ans =
    70    80    85    91    95    100
```

12.4.1 Indexing into Vectors of Structures

Often, when the data structure is a vector of structures, it is necessary to iterate through the vector in order by different fields. For example, for the *packages* vector defined previously, it may be necessary to iterate in order by the *cost*, or by the *price* fields. Rather than sorting the entire vector of structures based on these fields, it may be more efficient to index into the vector based on these fields, for example, to have an index vector based on *cost* and another based on *price*.

	packages					cost_ind			price_ind
	item_no	cost	price	code					
1	123	19.99	39.95	'g'	1	2	1	3
2	456	5.99	49.99	'l'	2	3	2	1
3	587	11.11	33.33	'w'	3	1	3	2

These index vectors would be created as before, comparing the fields but exchanging the values in the index vectors. Once the index vectors have been created, they can be used to iterate through the *packages* vector in the desired order. For example, the function to print the information from *packages* has been modified so that in addition to the vector of structures, the index vector is also passed, and the function iterates using that index vector.

printpackind.m

```
function printpackind(packstruct, indvec)
% This function prints a table showing all
% values from a vector of packages structures
fprintf('Item # Cost Price Code \n')
no_packs = length(packstruct);
for i = 1:no_packs
    fprintf('%6d %6.2f %6.2f %3c\n', ...
        packstruct(indvec(i)).item_no, ...
        packstruct(indvec(i)).cost, ...
        packstruct(indvec(i)).price, ...
        packstruct(indvec(i)).code)
end
```

```
>> printpackind(packages,cost_ind)
Item #   Cost      Price     Code
  456     5.99     49.99       l
  587    11.11     33.33       w
  123    19.99     39.95       g
>> printpackind(packages,price_ind)
Item #   Cost      Price     Code
  587    11.11     33.33       w
  123    19.99     39.95       g
  456     5.99     49.99       l
```

 PRACTICE 12.3

Modify the function *createind* to create the *cost_ind* index vector.

12.5 SEARCHING

Searching means looking for a value (a *key*) in a list or in a vector. We already have seen that MATLAB has a function, **find**, that will return the indices in an array that meet a criterion. To examine the programming methodologies, we will in this section examine two search algorithms:

- *sequential search*
- *binary search*

12.5.1 Sequential Search

A *sequential search* is accomplished by looping through the vector element-by-element starting from the beginning, looking for the key. Normally the index of the element in which the key is found is what is returned. For example, here is a function that will search a vector for a key and return the index or the value 0 if the key is not found:

seqsearch.m

```
function index = seqsearch(vec, key)
% Implements a sequential search through a
% vector looking for a key; returns the index
len = length(vec);
index = 0;
for i = 1:len
    if vec(i) == key
        index = i;
    end
end
```

Here are two examples of calling this function:

```
>> values = [85 70 100 95 80 91];
>> key = 95;
>> seqsearch(values, key)
ans =
      4
>> seqsearch(values, 77)
ans =
      0
```

This example assumes that the key is found only in one element in the vector. Also, although it works, it is not a very efficient algorithm. If the vector is large, and the key is found in the beginning, this still loops through the rest of the vector. An improved version would loop until the key is found or the entire vector has been searched. In other words, a **while** loop is used rather than a **for** loop; there are two parts to the condition.

smartseqsearch.m

```
function index = smartseqsearch(vec, key)
% Smart sequential search; searches a vector
% for a key but ends when it is found
% Returns the index or 0 if not found
len = length(vec);
index = 0;
i = 1;
while i < len && vec(i) ~= key
      i = i + 1;
end
if vec(i) == key
    index = i;
end
```

12.5.2 Binary Search

The *binary search* assumes that the vector has been sorted first. The algorithm is similar to the way it works when looking for a name in a phone directory (which is sorted alphabetically). To find the value of a key, look at the element in the middle.

- If that is the key, the index has been found.
- If it is not the key, decide whether to search the elements before or after this location and adjust the range of values in which the search is taking place and start this process again.

To implement this, we will use variables *low* and *high* to specify the range of values in which to search. To begin, the value of *low* will be 1, and the value of *high* will be the length of the vector. The variable *mid* will be the index of the element in the middle of the range from *low* to *high*. If the key is not found at *mid*, there are two possible ways to adjust the range. If the key is less than the value at *mid*, we change *high* to *mid* − 1. If the key is greater than the value at *mid*, we change *low* to *mid* + 1.

An example is to search for the key of 91 in the vector

1	2	3	4	5	6
70	80	85	91	95	100

The following table shows what will happen in each iteration of this search algorithm.

Iteration	Low	High	Mid	Found?	Action
1	1	6	3	No	Move low to mid + 1
2	4	6	5	No	Move high to mid − 1
3	4	4	4	Yes	Done! Index is mid

The key was found in the fourth element of the vector.

Another example: search for the key of 82:

Iteration	Low	High	Mid	Found?	Action
1	1	6	3	No	Move high to mid −1
2	1	2	1	No	Move low to mid + 1
3	2	2	2	No	Move low to mid + 1
4	3	2	→ this ends it!		

The value of *low* can't be greater than *high*; this means that the key is not in the vector. So, the algorithm repeats until either the key is found or until *low* > *high*, which means the key is not there.

The following function implements this binary search algorithm. The function receives two arguments: the sorted vector, and a key (alternatively, the function could sort the vector). The values of *low* and *high* are initialized to the first and last indices in the vector. The output argument *outind* is initialized to −1, which is the value that the function will return if the key is not found. The function loops until either *low* is greater than *high*, or until the key is found.

binsearch.m

```
function outind = binsearch(vec, key)
% This function implements a binary search
low = 1;
high = length(vec);
outind = -1;
while low <= high && outind == -1
    mid = floor((low + high)/2);
    if vec(mid) == key
      outind = mid;
    elseif key < vec(mid)
      high = mid - 1;
    else
      low = mid + 1;
    end
end
```

Here are examples of calling this function:

```
>> vec = randint(1,7, [1 30])
vec =
        2    11    25     1     5     7     6
>> svec = sort(vec)
svec =
        1     2     5     6     7    11    25
>> binsearch(svec, 4)
ans =
       -1
>> binsearch(svec, 25)
ans =
      7
>> binsearch(svec, 5)
ans =
      3
```

The binary search can also be implemented as a recursive function. The following recursive function also implements this binary search algorithm. The function receives four arguments: a sorted vector, a key to search for, and the values of *low* and *high* (which, to begin, will be 1 and the length of the vector). It will return −1 if the key is not in the vector, or the index of the element in which it is found. The base cases in the algorithm are when *low* > *high*, which means the key is not in the vector, or when it is found. Otherwise, the general case is to adjust the range and call the binary search function again.

recbinsearch.m

```
function outind = recbinsearch(vec, key, low, high)
% Recursive binary search function
mid = floor((low + high)/2);
if low > high
    outind = -1;
elseif vec(mid) == key
    outind = mid;
elseif key < vec(mid)
    outind = recbinsearch(vec,key,low,mid-1);
else
    outind = recbinsearch(vec,key,mid+1,high);
end
```

Here are examples of calling this function:

```
>> recbinsearch(svec, 5,1,length(svec))
ans =
     3
>> recbinsearch(svec, 25,1,length(svec))
ans =
     7
>> recbinsearch(svec, 4,1,length(svec))
ans =
    -1
```

SUMMARY

Common Pitfalls

- Forgetting that **max** and **min** return the index of only the first occurrence of the maximum or minimum value
- Not realizing that a data set has outliers that can drastically alter the results obtained from the statistical functions

- When sorting a vector of structures on a field, forgetting that although only the field in question is compared in the sort algorithm, the entire structures must be interchanged
- Forgetting that a data set must be sorted before using a binary search

Programming Style Guidelines

- Remove the largest and smallest numbers from a large data set before performing statistical analyses, in order to handle the problem of outliers.
- Use **sortrows** to sort strings stored in a matrix alphabetically; for cell arrays, **sort** can be used.
- When it is necessary to iterate through a vector of structures in order based on several different fields, it may be more efficient to create index vectors based on these fields rather than sorting the vector of structures multiple times.

MATLAB Functions and Commands			
mean	median	setdiff	sort
var	union	setxor	sortrows
std	intersect	ismember	
mode	unique	issorted	

Exercises

1. Experimental data values are stored in a file. Create a file in a matrix form with random values for testing. Write a script that will **load** the data, and then determine the difference between the largest and smallest numbers in the file.

2. The **range** of a data set is the difference between the largest value and the smallest. A data file 'tensile.dat' stores the tensile strength of some aluminum samples. Create a test data file; read in the tensile strengths and print the range.

3. Write a function *mymin* that will receive any number of arguments, and will return the minimum. For example,

```
>> mymin(3, 6, 77, 2, 99)
ans =
2
```

   Note: The function is not receiving a vector; rather, all the values are separate arguments.

4. The chemical balance of a swimming pool is important for the safety of the swimmers. The pH level of a pool has been measured every day and the results are stored in a file. Create a data file to simulate these measurements; the values should be random numbers in the range from 7 to 8. Read the pH values from the file and calculate the mean and standard deviation of the pH values.

5. A batch of 500-ohm resistors is being tested by a quality engineer. A file called 'testresist.dat' stores the resistance of some resistors that have been measured.

The resistances have been stored one per line in the file. Create a data file in this format. Then, load the information and calculate and print the mean, median, mode, and standard deviation of the resistances. Also, calculate how many of the resistors are within 1% of 500 ohms.

6. Write a function *calcvals* that will calculate the maximum, minimum, and mean value of a vector based on how many output arguments are used to call the function. Examples of function calls are as follows:

```
>> vec = [4 9 5 6 2 7 16 0];
>> [mmax, mmin, mmean] = calcvals(vec)
mmax =
       16
mmin =
       0
mmean =
       6
>> [mmax, mmin] = calcvals(vec)
mmax =
       16
mmin =
       0
>> mmax = calcvals(vec)
mmax =
       16
```

7. Write a script that will do the following. Create two vectors with 20 random integers in each; in one the integers should range from 1 to 5, and in the other from 1 to 500. For each vector, would you expect the mean and median to be approximately the same? Would you expect the standard deviation of the two vectors to be approximately the same? Answer these questions, and then use the built-in functions to find the minimum, maximum, mean, median, standard deviation, and mode of each. Do a histogram for each in a subplot. Run the script a few times to see the variations.

8. Write a function that will return the mean of the values in a vector, not including the minimum and maximum values. Assume that the values in the vector are unique. It is OK to use the built-in **mean** function. To test this, create a vector of 10 random integers, each in the range from 0 to 50, and pass this vector to the function.

9. Write a function *mymedian* that will receive a vector as an input argument, and will sort the vector and return the median. Any built-in functions may be used, except the **median** function. Loops may not be used. Here are some examples of function calls:

```
>> mymedian(1:5)
ans =
       3
```

```
>> mymedian(1:6)
ans =
      3.5000
>> mymedian([5 9 2])
ans =
    5
>> m = mymedian([5 9 2 4])
m =
    4.5000
```

10. What is the difference between the mean and the median of a data set if there are only two values in it?

11. A student missed one of four exams in a course, and the professor decided to use the average of the other three grades for the missed exam grade. Which would be better for the student: the mean or the median if the three recorded grades were 99, 88, and 95? What if the grades were 99, 70, and 77?

12. A weighted mean is used when there are varying weights for the data values. For a data set given by x = {x1, x2, x3, x4,..., xn} and corresponding weights for each xi, w = {w1, w2, w3, w4,..., wn}, the weighted mean is

$$\frac{\sum_{i=1}^{n} x_i w_i}{\sum_{i=1}^{n} w_i}.$$

For example, assume that in an economics course there are three quizzes and two exams, and the exams are weighted twice as much as the quizzes. If the quiz scores are 95, 70, and 80 and the exam scores are 85 and 90, the weighted mean would be:

$$\frac{95*1+70*1+80*1+85*2+90*2}{1+1+1+2+2} = \frac{595}{7} = 85$$

Write a function that will receive two vectors as input arguments: one for the data values and one for the weights, and will return the weighted mean.

13. A production facility is producing some nails that are supposed to have a diameter of 0.15 inch. At five different times, 10 sample nails were measured; their diameters were stored in a file that has five lines and 10 diameters on each. First, create a data file to simulate this data. Then, write a script to print the mean and standard deviation for each of the five sets of sample nails.

14. The **coefficient of variation** is useful when comparing data sets that have quite different means. The formula is CV = (standard deviation/mean) * 100%. A history course has two different sections; their final exam scores are stored in two separate rows in a file. For example,

```
99 100 95 92 98 89 72 95 100 100
83 85 77 62 68 84 91 59 60
```

Create this data file, read the data into vectors, and then use the CV to compare the two sections of this course.

15. Write a function *allparts* that will read in lists of part numbers for parts produced by two factories. These are contained in data files called 'xyparts.dat' and 'qzparts.dat'. The function will return a vector of all parts produced, in sorted order (with no repeats). For example, if the file 'xyparts.dat' contains

```
123 145 111 333 456 102
```

and the file 'qzparts.dat' contains

```
876 333 102 456 903 111
```

calling the function would return the following:

```
>> partslist = allparts
partslist =
```

```
102 111 123 145 333 456 876 903
```

16. The set functions can be used with cell arrays of strings. Create two cell arrays to store (as strings) course numbers taken by two students. For example,

```
s1 = {'EC 101', 'CH 100', 'MA 115'};
s2 = {'CH 100', 'MA 112', 'BI 101'};
```

Use a set function to determine which courses the students have in common.

17. A vector v is supposed to store unique random numbers. Use set functions to determine whether or not this is true.

18. Write a function *mydsort* that sorts a vector in descending order (using a loop, not the built-in **sort** function).

19. In product design, it is useful to gauge how important different features of the product would be to potential customers. One method of determining which features are most important is a survey in which people are asked for different features: "Is this feature important to you?" The number of potential customers who responded Yes is then tallied. For example, a company conducted such a survey for 10 different features; 200 people took part in the survey. The data was collected into a file that might look like this:

1	2	3	4	5	6	7	8	9	10
30	83	167	21	45	56	55	129	69	55

A **Pareto chart** is a bar chart in which the bars are arranged in decreasing values. The bars on the left in a Pareto chart indicate which are the most important features. Create a data file, and then a **subplot** to display the data with a bar chart organized by question on the left and a Pareto chart on the right.

20. DNA is a double-stranded helical polymer that contains basic genetic information in the form of patterns of nucleotide bases. The patterns of the base molecules A, T,

C, and G encode the genetic information. Construct a cell array to store some DNA sequences as strings, such as

```
TACGGCAT
ACCGTAC
```

and then sort these alphabetically. Next, construct a matrix to store some DNA sequences of the same length and then sort them alphabetically.

21. Write a function *matsort* to sort all the values in a matrix (decide whether the sorted values are stored by row or by column). It will receive one matrix argument and return a sorted matrix. Do this without loops, using the built-in functions **sort** and **reshape**. For example:

```
>> mat
mat =

     4     5     2
     1     3     6
     7     8     4
     9     1     5
>> matsort(mat)
ans =

     1     4     6
     1     4     7
     2     5     8
     3     5     9
```

22. Write a function that will receive two arguments: a vector and a character (either 'a' or 'd') and will sort the vector in the order specified by the character (ascending or descending).

23. Write a function that will receive a vector and will return two index vectors: one for ascending order, and one for descending order. Check the function by writing a script that will call the function and then use the index vectors to print the original vector in ascending and descending order.

24. Write a function *myfind* that will search for a key in a vector and return the indices of all occurrences of the key, like the built-in **find** function. It will receive two arguments: the vector and the key, and will return a vector of indices (or the empty vector [] if the key is not found).

CHAPTER 13

Sights and Sounds

KEY WORDS

sound signal
sampling frequency
audio file format
graphics primitive
object-oriented
  programming

parent/child
core object
text box
pixel
true color
RGB

colormap
Graphical User
  Interface
event
callback function

CONTENTS

The MATLAB® product has functions that manipulate audio or sound files and also images. This chapter will start with a brief introduction to some of the sound processing functions. We have seen that MATLAB uses Handle Graphics® in its figures. Some of the core graphics objects used to create figures will be explored. Image processing functions will be introduced, and the two basic methods for representing color in images will be explained. Finally, this chapter will introduce the topic of Graphical User Interfaces from a programming standpoint.

13.1 SOUND FILES

A *sound signal* is an example of a continuous signal that is sampled to result in a discrete signal. In this case, sound waves traveling through the air are recorded as a set of measurements that can then be used to reconstruct the original sound signal as closely as possible. The sampling rate or *sampling frequency* is the number of samples taken per time unit, for example per second. Sound signals are usually measured in Hertz (Hz).

In MATLAB, the discrete sound signal is represented by a vector, and the frequency is measured in Hertz. MATLAB has several MAT-files that store, for various sounds, the signal vector in a variable *y* and the frequency in a variable *Fs*. These MAT-files include **chirp**, **gong**, **laughter**, **splat**, **train**, and **handel**. There is a built-in function, **sound**, that will send a sound signal to an output device such as speakers. The function call

```
>> sound(y,Fs)
```

will play the sound represented by the vector *y* at the frequency *Fs*. For example, to hear a gong, load the variables from the MAT-file and then play the sound using the **sound** function:

```
>> load gong
>> sound(y,Fs)
```

Sound is actually a wave; the amplitudes are what are stored in the sound signal variable *y*. These are supposed to be in the range from –1 to 1. The **plot** function can be used to display the data. For example, the following script creates a **subplot** that displays the signals from **chirp** and from **train**, as seen in Figure 13.1.

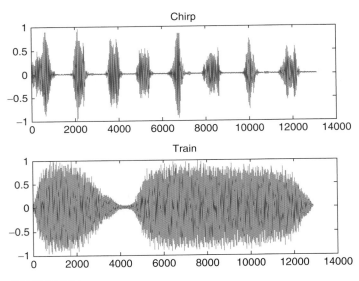

FIGURE 13.1

Amplitudes from chirp and train.

chirptrain.m

```
% Display the sound signals from chirp and train
load chirp
subplot(2,1,1)
plot(y)
title('Chirp')
load train
subplot(2,1,2)
plot(y)
title('Train')
```

The first argument to the **sound** function can be an n × 2 matrix for stereo sound. Also, the second argument can be omitted when calling the **sound** function, in which case the default sample frequency of 8192 Hz is used. This is the frequency stored in the built-in sound MAT-files.

```
>> load train
Fs
Fs =
      8192
```

MATLAB has several other functions that let you read and play sound or audio files. In audio files, sampled data for each audio channel is stored. There are several *audio file formats* that are used in industry, on different computer platforms. Audio files with the extension .au were developed by Sun Microsystems, and typically are used with Java and UNIX, whereas Windows PCs typically use .wav files that were developed by Microsoft.

> **PRACTICE 13.1**
>
> If you have speakers, try loading one of the sound MAT-files, and use the **sound** function to play the sound. Then, change the frequency; for example, multiply the variable Fs by 2 and by 0.5 and play these sounds again.
>
> ```
> >> load train
> >> sound(y, Fs)
> >> sound(y, Fs*2)
> >> sound(y, Fs*.5)
> ```

MATLAB has functions **wavread** that will read a .wav file, **wavrecord** that will record, **wavwrite** that will write a sound file, and **wavplay** that will play one. The default frequency for these functions is 11025 Hz.

For .au files, there are functions **auread** to read and **auwrite** to write in this format.

13.2 INTRODUCTION TO HANDLE GRAPHICS

With Handle Graphics, all figures consist of different *objects*, each of which is assigned a *handle*. The object handle is a unique real number that is used to refer to the object.

13.2.1 Graphics Objects and Their Properties

Objects include *graphics primitives* such as lines and text, as well as the axes used to orient the objects. The objects are organized hierarchically, and there are properties associated with each object. This is the basis of *object-oriented programming*: objects are organized hierarchically (e.g., a *parent* comes before its *children* in the hierarchy) and this hierarchy has ramifications in terms of the properties; generally children inherit properties from the parents. The hierarchy in MATLAB can be seen in the Help, "Organization of Graphics Objects." In this, it can be seen that the organization includes

In other words, the Figure Window includes Axes, which are used to orient Core objects (primitives such as line, rectangle, text) and Plot objects (which are used to produce bar charts, area plots, etc.).

We have seen already that once an object handle has been stored in a variable, the function **get** can be used to find its properties, and **set** can be used to change them.

13.2.1.1 Core Objects

The *Core Objects* in MATLAB are the very basic graphics primitives. A description can be found under the MATLAB Help: Under the Contents tab, click Handle Graphics Objects, and then Core Graphics Objects. The core objects include:

- **line**
- **text**
- **rectangle**
- **patch**
- **image**

These are all built-in functions; **help** can be used to find out how each function is used.

One core graphics object is a **line**, which is also what the **plot** function produces. Here is an example of creating a line object, modifying some properties, and saving the handle in a variable *hl*:

```
>> x = -2*pi: 1/5 : 2*pi;
>> y = sin(x);
>> hl = line(x,y,'LineWidth', 3, 'Color', [0.5 0.5 0.5])
hl =
    159.0405
```

As seen in Figure 13.2, this draws a reasonably thick grey line for the **sin** function. As before, the handle will be valid as long as the Figure Window is not closed. Some of the properties of this object are:

```
>> get(hl)
    Color = [0.5 0.5 0.5]
    EraseMode = normal
    LineStyle = -
```

```
LineWidth = [3]
Marker = none
MarkerSize = [6]
t = auto
MarkerFaceColor = none
XData = [ (1 by 63) double array]
YData = [ (1 by 63) double array]
ZData = []
        etc.
```

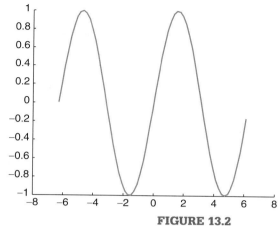

FIGURE 13.2

*A **line** object with modified line width and color.*

The **text** graphics function allows text to be printed in a Figure Window, including special characters that are printed using \specchar, where specchar is the actual name of the special character. The format of a call to the **text** function is

```
text(x,y,'text string')
```

where x and y are the coordinates on the graph of the beginning of the **text box** in which the text string appears.

To see the options for the special characters, under the Contents tab in Help, click Handle Graphics Property Browser, then click Core Objects, and then choose Text. The special characters are shown in a table under the String property. The special characters include letters of the Greek alphabet, arrows, and characters frequently used in equations. For example, Figure 13.3 displays the Greek symbol for pi and a right arrow within the text box.

```
>> x = -4:0.2:4;
>> y = sin(x);
>> hp = line(x,y,'LineWidth',3);
>> thand = text(2,0,'Sin(\pi)\rightarrow')
```

FIGURE 13.3

*A **line** object with a text box.*

Using **get** will display properties of the text box, for example:

```
>> get(thand)
  BackgroundColor = none
  Color = [0 0 0]
  EdgeColor = none
  EraseMode = normal
  Editing = off
  Extent = [1.95862 -0.0670554 0.901149
    0.110787]
  FontAngle = normal
  FontName = Helvetica
  FontSize = [10]
```

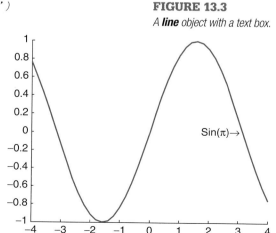

```
FontUnits = points
FontWeight = normal
HorizontalAlignment = left
LineStyle = -
LineWidth = [0.5]
Margin = [2]
Position = [2 0 0]
Rotation = [0]
String = Sin(\pi)\rightarrow
Units = data
Interpreter = tex
VerticalAlignment = middle
   etc.
```

FIGURE 13.4

Text box with a modified edge color and background color.

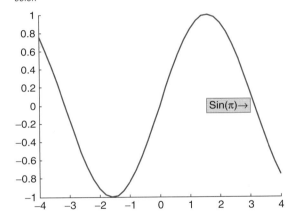

Although the Position specified was (2,0), the Extent is the actual extent of the text box, which cannot be seen since the BackgroundColor and EdgeColor are not specified. These can be changed using **set**. For example, the following produces the result seen in Figure 13.4:

```
>> set(thand,'BackgroundColor',[0.8 0.8 0.8],...
       'EdgeColor',[1 0 0])
```

When the Units property has the value of *data*, which is the default as shown, the Extent of the text box is given by a vector [x y width height] where x and y are the coordinates of the bottom left-hand corner of the text box.

Another core graphics object is the **rectangle**, which can have curvature added to it (!!). Just calling the function **rectangle** without any arguments brings up a Figure Window (seen in Figure 13.5) which at first glance doesn't seem to have anything in it:

```
>> recthand = rectangle;
```

Using the **get** function will display the properties, some of which are excerpted here:

```
>> get(recthand)
        Curvature = [0 0]
        FaceColor = none
        EdgeColor = [0 0 0]
        LineStyle = -
        LineWidth = [0.5]
```

FIGURE 13.5

*A **rectangle** object.*

```
Position = [0 0 1 1]
Type = rectangle
```

The Position of a rectangle is [x y w h], where x and y are the coordinates of the lower left point, w is the width, and h is the height. The default rectangle has a Position of [0 0 1 1]. The default Curvature is [0 0], which means no curvature. The values range from [0 0] (no curvature) to [1 1] (ellipse). A more interesting rectangle object is seen in Figure 13.6. Notice that properties can be set when calling the **rectangle** function, and also subsequently using the **set** function.

```
>> rh = rectangle('Position', [0.2, 0.2, 0.5, 0.8],...
   'Curvature',[0.5, 0.5]);
>> axis([0 1.2 0 1.2])
>> set(rh,'Linewidth',3,'LineStyle',':')
```

FIGURE 13.6
Rectangle object with curvature.

The **patch** function is used to create a patch graphics object, which is made from two-dimensional polygons. The patch is defined by its vertices and faces. For example, consider a patch that has four vertices in three-dimensional space, given by the coordinates:

```
(0, 0, 0)
(1, 0, 0)
(0, 1, 0)
(0.5, 0.5, 1)
```

Just to visualize these points first, the **plot3** function can be used, as seen in Figure 13.7.

```
>> x = [0 1 0 0.5];
>> y = [0 0 1 0.5];
>> z = [0 0 0 1];
>> plot3(x,y,z,'ko')
```

FIGURE 13.7
*Visualization of three-dimensional points using **plot3**.*

Rotating the figure displays all four points, as seen in Figure 13.8.

Creating and displaying these points is not necessary to create a patch object; this was just done to visualize the vertices first. A patch object is defined by both the vertices and the faces of the polygon that connect these vertices. One way of calling this function is patch(fv), where *fv* is a structure variable with fields called *vertices* and *faces*.

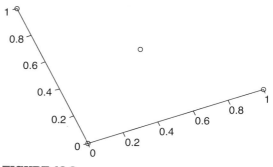

FIGURE 13.8
Three-dimensional points rotated.

FIGURE 13.9
Patch object.

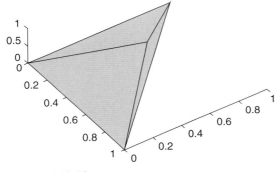

FIGURE 13.10
*Rotated **patch** object.*

```
polyhedron.vertices = [...
0 0 0
1 0 0
0 1 0
0.5 0.5 1];
polyhedron.faces = [...
1 2 3
1 2 4
1 3 4
2 3 4];
pobj = patch(polyhedron, ...
'FaceColor',[0.8, 0.8, 0.8],...
'EdgeColor','black');
```

The *polyhedron.vertices* field is a matrix in which each row represents (x,y,z) points. The field *polyhedron.faces* defines the faces: for example the first row in the matrix specifies to draw a line from vertex 1 to vertex 2 to vertex 3 to form the first face. The face color is set to grey and the edge color to black. The figure, as seen in Figure 13.9, shows only two faces. Using the rotate icon on the Figure Window, the figure can be rotated to see the other edges as seen in Figure 13.10.

13.3 IMAGE PROCESSING

Images are represented as grids, or matrices, of picture elements (called **pixels**). In MATLAB an image typically is represented as a matrix in which each element corresponds to a pixel in the image. Each element that represents a particular pixel stores the color for that pixel. There are two basic ways that the color can be represented:

- *True color*, or **RGB**, in which the three color components are stored (red, green, and blue, in that order).

- Index into a *colormap*: the value stored is an integer that refers to a row in a matrix called a colormap. The colormap stores the red, green, and blue components in three separate columns.

For an image that has m × n pixels, the true color matrix would be a three-dimensional matrix with the size m × n × 3. The first two dimensions represent the coordinates of the pixel. The third index is the color component; (:,:,1) is the red, (:,:,2) is the green, and (:,:,3) is the blue component.

The indexed representation instead would be an m × n matrix of integers, each of which is an index into a colormap matrix that is the size $p \times 3$ (where p is the number of colors available in that particular colormap). Each row in the colormap has three numbers representing one color: first the red, then green, then blue components, as we have seen before. For example,

```
[1 0 0] is red
[0 1 0] is green
[0 0 1] is blue
     etc.
```

MATLAB has several built-in colormaps that are named; the reference page on **colormap** displays them. Calling the function **colormap** without passing any arguments will return the current colormap, which by default is one called *jet*.

The following stores the current colormap in a variable *map*, gets the size of the matrix (which will be the number of rows in this matrix, or in other words the number of colors, by 3 columns), and displays the first five rows in this colormap. If the current colormap is the default *jet*, this will be the result:

```
>> map = colormap;
>> [r c] = size(map)
r =
     64
c =
     3
>> map(1:5,:)
ans =

     0          0          0.5625
     0          0          0.6250
     0          0          0.6875
     0          0          0.7500
     0          0          0.8125
```

This shows that there are 64 rows, or in other words, 64 colors, in this particular colormap. It also shows that the first five colors are shades of blue.

The format of calling the **image** function is:

```
image(mat)
```

where the matrix *mat* is a matrix that represents the colors in an m × n image (m × n pixels in the image). If the matrix has the size m × n, then each element is an index into the current colormap.

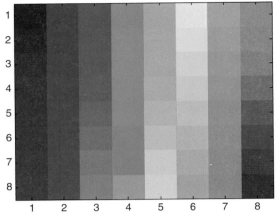

FIGURE 13.11

Columnwise display of the 64 colors in the jet colormap.

One way to display the colors in the *jet* colormap (which has 64 colors) is to create a matrix that stores the values 1 through 64, and pass that to the **image** function, as seen in Figure 13.11. When the matrix is passed to the **image** function, the value in each element in the matrix is used as an index into the colormap. For example, the value in cmap(1,2) is 9, so the color displayed in location (1,2) in the image will be the color represented by the ninth row in the colormap. By using the numbers 1 through 64, we can see all the colors in this colormap.

```
>> cmap = reshape(1:64, 8,8)
cmap =

    1     9    17    25    33    41    49    57
    2    10    18    26    34    42    50    58
    3    11    19    27    35    43    51    59
    4    12    20    28    36    44    52    60
    5    13    21    29    37    45    53    61
    6    14    22    30    38    46    54    62
    7    15    23    31    39    47    55    63
    8    16    24    32    40    48    56    64

>> image(cmap)
```

This shows that the first colors are shades of blue, the last colors are shades of red, and in between there are shades of aqua, green, yellow, and orange.

FIGURE 13.12

A 5 × 5 display of random colors from the jet colormap.

Another example creates a 5 × 5 matrix of random integers in the range from 1 to the number of colors; the resulting image is seen in Figure 13.12.

```
>> mat = randint(5,5,[1, r])
mat =

    54    33    13    45    32
     2    46    44    25    58
    44    28    20    56    53
    25    20    35    55    42
    54    13    10    38    53

>> image(mat)
```

Of course, these images are rather crude; the elements representing the pixel colors are quite large blocks. A larger matrix would result in something more closely resembling an image, as displayed in Figure 13.13.

```
>> mat = randint(500,500,[1, r]);
>> image(mat)
```

Although MATLAB has built-in colormaps, it is also possible to create others using combinations of any colors. For example, the following creates a customized colormap with just three colors: black, white, and red. This is then set to be the current colormap by passing the color map matrix to the **colormap** function. Then, a 40 × 40 matrix of random integers in the range from 1 to 3 (since there are just three colors) is created, and that is passed to the **image** function; the results are seen in Figure 13.14.

FIGURE 13.13
A 500 × 500 display of random colors.

```
>> mycolormap = [0 0 0; 1 1 1; 1 0 0]
mycolormap =

     0     0     0
     1     1     1
     1     0     0

>> colormap(mycolormap)
>> mat = randint(40,40,[1 3]);
>> image(mat)
```

True color matrices are another way to represent images. True color matrices are three-dimensional matrices. The first two coordinates are the coordinates of the pixel. The third index is the color component; (:,:,1) is the red, (:,:,2) is the green, and (:,:,3) is the blue component. Each element in the matrix is of the type **uint8**, which is an unsigned integer type using 8 bits (unsigned means that the + or − sign is not stored). With 8 bits, 2^8 (or 256) values can be stored and so the range of values is 0

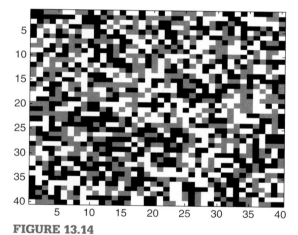

FIGURE 13.14
Random colors from a custom colormap.

to 255. The minimum value, 0, represents the darkest hue available so all 0's results in a black pixel. The maximum value, 255, represents the brightest hue. For example, if the values for a given pixel coordinates px and py (px,py,1) is 255, (px,py,2) is 0 and (px,py,3) is 0 then that pixel will be bright red. All 255's results in a white pixel.

The **image** function displays the information in the three-dimensional matrix as an image.

For example, this creates a 2 × 2 image as seen in Figure 13.15. The matrix is 2 × 2 × 3, where the third dimension is the color. The pixel in location (1,1) is red,

FIGURE 13.15

Image from a true color matrix.

the pixel in location (1,2) is blue, the pixel in location (2,1) is green, and the pixel in location (2,2) is black.

```
>> mat(1,1,1) = 255;
>> mat(1,1,2) = 0;
>> mat(1,1,3) = 0;
>> mat(1,2,1) = 0;
>> mat(1,2,2) = 0;
>> mat(1,2,3) = 255;
>> mat(2,1,1) = 0;
>> mat(2,1,2) = 255;
>> mat(2,1,3) = 0;
>> mat(2,2,1) = 0;
>> mat(2,2,2) = 0;
>> mat(2,2,3) = 0;
>> mat = uint8(mat);
>> image(mat)
```

The function **imread** can read in an image file, for example a JPEG (.jpg) file. The function reads color images into a three-dimensional matrix.

```
>> myimage1 = imread('Fishing_1.JPG');
>> size(myimage1)
ans =
        1536    2048       3
```

FIGURE 13.16

*Image from a JPEG file displayed using **image**.*

In this case, the image is represented as true color matrix. This indicates that the image has 1536 × 2048 pixels. The **image** function displays the information in this three-dimensional matrix as an image, as seen in Figure 13.16.

```
>> image(myimage1)
```

The image can be changed by manipulating the numbers in the matrix. For example, multiplying every number by 0.75 will result in a range of values from 0 to 191 instead of from 0 to 255. Since the larger numbers are brighter, this has the effect of dimming the hues in the pixels, as seen in Figure 13.17.

```
>> dimmer = 0.75*myimage1;
>> image(dimmer)
```

13.4 INTRODUCTION TO GRAPHICAL USER INTERFACES

Graphical user interfaces, or GUIs, are essentially objects that allow users to have input using graphical interfaces such as push buttons, sliders, radio buttons, toggle buttons, pop-up menus, and so forth. GUIs are an example of object-oriented programming in which there is a hierarchy. For example, the parent may be a Figure Window and its children would be graphics objects such as push buttons and text boxes.

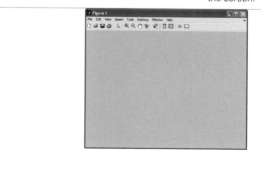

FIGURE 13.17
Image dimmed by manipulating the matrix.

The parent user interface object can be a **figure**, **uipanel**, or **uibuttongroup**. A **figure** is a Figure Window created by the **figure** function. A **uipanel** is a means of grouping together user interface objects (the **ui** stands for user interface). A **uibuttongroup** is a means of grouping together buttons (both radio buttons and toggle buttons).

In MATLAB there are two basic methods for creating GUIs: writing the GUI program from scratch, or using the built-in Graphical User Interface Development Environment (GUIDE). GUIDE allows the user to graphically lay out the GUI, and then MATLAB generates the code for it automatically. However, in order to be able to understand and modify this code, it is important to understand the underlying programming concepts. Therefore, this section will concentrate on the programming methodology.

A Figure Window is the parent of any GUI. Just calling the **figure** function will bring up a blank Figure Window. Assigning the handle of this Figure Window to a variable and then using the **get** function will show the default properties. These properties, such as the color of the window, its position on the screen, and so forth can be changed using the **set** function or when calling the **figure** function to begin with. For example,

```
>> f = figure;
```

brings up a grey figure box near the top of the screen as seen in Figure 13.18.

Some of its properties are excerpted here:

FIGURE 13.18
*Placement of **figure** within the screen.*

```
>> get(f)
  Color = [0.8 0.8 0.8]
  Colormap = [ (64 by 3) double array]
  Position = [360 502 560 420]
  Units = pixels
  Children = []
  Visible = on
```

The position vector specifies [left bottom width height]. The first two numbers, the left and bottom, are the distance that the lower left-hand corner of the figure box is from the lower left of the monitor screen (first from the left and then from the bottom). The last two are the width and height of the figure box itself. All these are in the default units of pixels.

The Visible property *on* means that the Figure Window can be seen. When creating a GUI, however, the normal procedure is to create the parent Figure Window but make it invisible. Then, all user interface objects are added to it, and properties are set. When everything has been completed, the GUI is made visible.

Most user interface objects are created using the **uicontrol** function. The Style property defines the type of object, as a string. For example, 'text' is the Style of a static text box, which normally is used as a label for other objects in the GUI, or for instructions.

The following example creates a GUI that just consists of a static text box in a Figure Window. The figure is first created but made invisible. The color is white, and it is given a position. Storing the handle of this figure in a variable allows the script to refer to it later on, to set properties, for example. The **uicontrol** function is used to create a text box, position it (the vector specifies the [left bottom width height] within the Figure Window itself), and put a string in it. Note that the position is within the Figure Window, not within the screen. A name is put on the top of the figure. The **movegui** function moves the GUI (the figure) to the center of the screen. Finally, when everything has been completed, the GUI is made visible. The **end** is at the end of the function because in the next examples, we will have nested functions that require **end** statements.

```
simple_gui.m
```

```
function simple_gui
% This is a simple GUI that just has a static text box

% Create the GUI but make it invisible for now while
% it is being initialized
f = figure('Visible', 'off','color','white','Position',...
     [300, 400, 450,250]);
```

(Continued)

```
htext = uicontrol('Style','text','Position', ...
    [200,50, 100, 25], 'String','My First GUI string');
% Put a name on it and move to the center of the screen
set(f,'Name','Simple GUI')
movegui(f,'center')
% Now the GUI is made visible
set(f,'Visible','on');
end
```

The Figure Window shown in Figure 13.19 will appear in the middle of the screen.

The static text box requires no interaction with the user.

In the next example, we will allow the user to enter a string in an editable text box, and then the GUI will print the user's string in red. In this example, there will be user interaction. First the user must type a string, and once this happens the user's entry in the editable text box will no longer be shown but instead the string that the user typed will be displayed in a larger red font, in a static text box. When the user's action (which is called an *event*) causes a response, a *call back function* is invoked. This is a *nested function* within the GUI function. Recall that when a function contains another function, they both must have an **end** statement. The algorithm for this example is:

- Create the Figure Window, but make it invisible.

- Make the color of the figure white, put a title on it, and move it to the center.

- Create a static text box with an instruction to enter a string.

- Create an editable text box:

  - the Style of this is 'edit'.

  - the call back function must be specified since the user's entry of a string necessitates a response (the function handle of the nested function is used).

- Make the GUI visible so the user can see the instruction and type a string.

- When the string is entered, the call back function *callbackfn* is called. Note that in the function header, there are two input arguments, *source* and *eventdata*. The input

FIGURE 13.19
Simple GUI with a static text box.

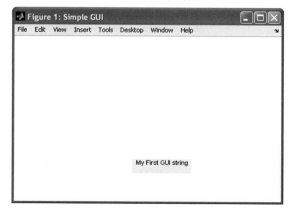

argument *source* refers to the uicontrol object that called it; *eventdata* is empty now (it may be used in MATLAB in a future version).

- The algorithm for the nested function *callbackfn* is:
    - make the previous GUI objects invisible.
    - get the string that the user typed (**Note:** either *source* or the function handle name *huitext* can be used to refer to the object in which the string was entered).
    - create a static text box to print the string in red with a larger font.
    - make this new object visible.

gui_with_editbox.m

```
function gui_with_editbox
% This is a GUI that has an editable text box
% and a callback function that prints the user's
% string in red

% Create the GUI but make it invisible for now
f = figure('Visible', 'off','color','white','Position',...
    [360, 500, 800,600]);
% Put a name on it and move it to the center of the screen
set(f,'Name','GUI with editable text')
movegui(f,'center')
% Create two objects: a box where the user can type and
% edit a string and also a text title for the edit box
hsttext = uicontrol('Style','text',...
    'BackgroundColor','white',...
    'Position',[100,425,400, 55],...
    'String','Enter your string here');
huitext = uicontrol('Style','edit',...
    'Position',[100,400,400,40],...
    'Callback',@callbackfn);
% Now the GUI is made visible
set(f,'Visible','on');
    % Call back function
```

(Continued)

```
function callbackfn(source,eventdata)
    set([hsttext huitext],'Visible','off');
    % Get the string that the user entered and print
    % it in big red letters
    printstr = get(huitext,'String');
    hstr = uicontrol('Style','text',...
    'BackgroundColor','white',...
    'Position',[100,400,400,55],...
    'String',printstr,...
    'ForegroundColor','Red','FontSize',30);
    set(hstr,'Visible','on')
    end
end
```

When the Figure Window first is made visible, the static text and the editable text box are shown. In this case, the user entered "hi and how are you?"

Note: In order to enter the string, the user must first click the mouse in the editable text box. The string that was entered by the user is seen in Figure 13.20.

Once the Enter key was pressed, the call back function was executed; the results are seen in Figure 13.21.

Now, we'll add a push button to the GUI. This time, the user will enter a string but the call back will be invoked when the push button is pushed.

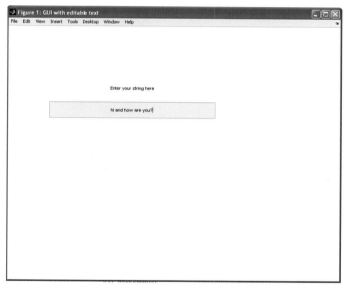

FIGURE 13.20
String entered by user in editable text box.

gui_with_pushbutton.m

```
function gui_with_pushbutton
% This is a GUI with an editable text box and a pushbutton

% Create the GUI but make it invisible for now while
```

(*Continued*)

```
% it is being initialized
f = figure('Visible',
'off','color','white','Position',...
    [360, 500, 800,600]);
hsttext = uicontrol('Style','text','BackgroundColor','white',...
    'Position',[100,425,400, 55],...
    'String','Enter your string here');
huitext = uicontrol('Style','edit','Position', [100,400,400,40]);
set(f,'Name','GUI with pushbutton')
movegui(f,'center')
% Create a pushbutton that says Push me!!
hbutton = uicontrol('Style','pushbutton','String',...
    'Push me!!', 'Position',[600,50,150,50], ...
    'Callback',@callbackfn);
% Now the GUI is made visible
set(f,'Visible','on');
    % Call back function
        function callbackfn(source,eventdata)
            set([hsttext huitext hbutton],'Visible','off');
            printstr = get(huitext,'String');
            hstr = uicontrol('Style','text','BackgroundColor',...
            'white', 'Position',[100,400,400,55],...
            'String',printstr, ...
            'ForegroundColor','Red','FontSize',30);
            set(hstr,'Visible','on')
        end
end
```

In this case the user types the string into the edit box. Pressing Enter, however, does not cause the call back function to be called; instead, the user must press the button with the mouse. The call back function is associated with the push button object. So, pushing the button will bring up the string in a larger red font. The push button is seen in Figure 13.22.

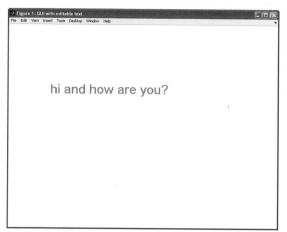

FIGURE 13.21

The result from the call back function.

FIGURE 13.22

GUI with a push button.

 PRACTICE 13.2

Create a GUI that will convert a length from inches to centimeters. The GUI should have an editable text box in which the user enters a length in inches, and a push button that says "Convert me!". Pushing the button causes the GUI to calculate the length in centimeters and display that. The call back function that accomplishes this should leave all objects visible. That means that the user can continue converting lengths until the Figure Window is closed. The GUI should display a default length to begin with (e.g., 1 inch). For example, calling the function might bring up the Figure Window shown in Figure 13.23.

Then, when the user enters a length (e.g., 5.2) and pushes the button, the Figure Window will show the new calculated length in centimeters (as seen in Figure 13.24).

FIGURE 13.23

Length conversion GUI with push button.

FIGURE 13.24

Result from conversion GUI.

Another GUI object that can be created is a slider. The slider object has a numerical value, and can be controlled by either clicking the arrows to move the value up or down, or by sliding the bar with the mouse. By default the numerical value ranges from 0 to 1, but these values can be modified using the Min and Max properties.

The function *gui_slider* creates in a Figure Window a slider that has a minimum value of 0 and a maximum value of 5. It uses text boxes to show the minimum and maximum values, and also the current value of the slider.

gui_slider.m

```
function gui_slider
% This is a GUI with a slider
f = figure('Visible', ...
'off','color','white','Position',...
[360, 500, 300,300]);
% Minimum and maximum values for slider
minval = 0;
maxval = 5;
% Create the slider object
slhan = uicontrol('Style','slider','Position',...
    [80,170,100, 50], ...
    'Min', minval, 'Max', maxval,'Callback', @callbackfn);
% Text boxes to show the minimum and maximum values
hmintext = uicontrol('Style','text','BackgroundColor','white', ...
    'Position', [40, 175, 30,30], 'String', num2str(minval));
hmaxtext = uicontrol('Style', ...
    'text','BackgroundColor','white',...
    'Position', [190, 175, 30,30], 'String', num2str(maxval));
% Text box to show the current value (off for now)
hsttext = uicontrol('Style','text','BackgroundColor','white',...
    'Position',[120,100,40,40],'Visible', 'off');
set(f,'Name','Slider Example')
movegui(f,'center')
set(f,'Visible','on');
    % Call back function displays the current slider value
```

(Continued)

```
function callbackfn(source,eventdata)
  num = get(slhan, 'Value');
  set(hsttext,'Visible','on','String',num2str(num))
end
end
```

Calling the function brings up the initial configuration seen in Figure 13.25.

Then, when the user interacts by sliding the bar or clicking an arrow, the current value of the slider is shown under it, as seen in Figure 13.26.

 PRACTICE 13.3

Use the Help browser to find the property that controls the increment value on the slider, and modify the *gui_slider* function to move in increments of 0.5 regardless of whether the arrows or slider is used.

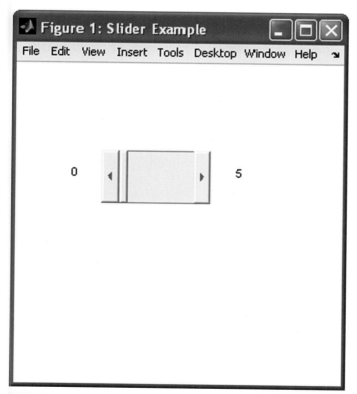

FIGURE 13.25
GUI with slider.

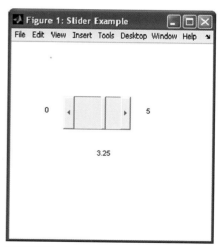

FIGURE 13.26
GUI with slider result shown.

SUMMARY

Common Pitfalls
- Confusing true color and colormap images
- Forgetting that **uicontrol** object positions are within the Figure Window, not within the screen

Programming Style Guidelines
- Make a GUI invisible while it is being created, so that everything becomes visible at once.

MATLAB Functions and Commands			
chirp	sound	text	uipanel
gong	wavread	rectangle	uibuttongroup
laughter	wavrecord	patch	uicontrol
splat	wavwrite	colormap	movegui
train	wavplay	image	
handel	line	imread	

Exercises

1. Load two of the built-in MAT-file sound files (e.g., **gong** and **chirp**). Store the sound vectors in two separate variables. Determine how to concatenate these so that the **sound** function will play one immediately followed by the other; for example, fill in the blank here:

   ```
   sound(  , 8192)
   ```

2. The function *playsound* (shown here) plays one of the built-in sounds. The function has a cell array that stores the names. When the function is called, an integer is passed that is an index into this cell array indicating which sound is to be played. The default is **train**, so if the user passes an invalid index that is used, the appropriate MAT-file is loaded. If the user passes a second argument, it is the frequency at which the sound should be played (otherwise, the default frequency is used). The function prints what sound is about to be played and at which frequency, and then actually plays this sound. You are to fill in the rest of the function. Here are examples of calling it (you can't hear it here, but the sound will be played!).

   ```
   >> playsound(-4)
   You are about to hear train at frequency 8192.0
   >> playsound(2)
   You are about to hear gong at frequency 8192.0
   >> playsound(3,8000)
   You are about to hear laughter at frequency 8000.0
   ```

playsound.m

```
function playsound(caind3, varargin)
%This function plays a sound
soundarray = {'chirp','gong','laughter','splat','train'};
if caind < 1 || caind > length(soundarray)
  caind = length(soundarray);
end
mysound = soundarray{caind};
eval(['load' mysound])
% Fill in the rest
```

3. In the MATLAB Help, under the Contents tab, click Functions by Category, then Graphics, then Handle Graphics, then text to get the MATLAB Function Reference on the function **text** (this is a lot more useful than just typing **help text**!). Read through this, and then on the very bottom click Text Properties for property descriptions. Create a graph, and then use the text function to put some text on it, including some \specchar commands to increase the font size and to print some Greek letters and symbols.

4. The cost of producing widgets includes an initial set-up cost plus an additional cost for each widget, so the total production cost per widget decreases as the number of widgets produced increases. The total revenue is a given dollar amount for each widget sold, so the revenue increases as the number sold increases. The break-even point is the number of widgets produced and sold for which the total production cost is equal to the total revenue. The production cost might be $5000 plus $3.55 per widget, and the widgets might sell for $10 each. Write a script that will find the break-even point using **solve**, and then plot the production cost and revenue functions on one graph for 1 to 1000 widgets. Print the break-even point on the graph using **text**.

5. Create a **rectangle** object, and use the **axis** function to change the axes so that you can see the rectangle easily. Change the Position, Curvature, EdgeColor, LineStyle, and LineWidth. Experiment with different values for the Curvature.

6. Write a script that will create the rectangle (shown in Figure 13.27) with a curved rectangle inside it and text inside that. The axes and dimensions of the Figure Window should be as shown here (you should approximate locations based on the axes shown in this figure). The font size for the string is 20. The curvature of the inner rectangle is [0.5, 0.5].

FIGURE 13.27
Nested rectangles with text box.

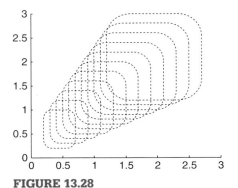

FIGURE 13.28

Curved rectangles produced in a loop.

7. Write a script that will start with a rounded rectangle. Change both the x and y axes from the default to go from 0 to 3. In a for loop, change the position vector by adding 0.1 to all elements 10 times (this will change the location and size of the rectangle each time). Create a movie consisting of the resulting rectangles. The final result should look like the plot shown in Figure 13.28.

8. Write a script that will create a two-dimensional **patch** object with just three vertices and one face connecting them. The x- and y-coordinates of the three vertices will be random real numbers in the range from 0 to 1. The lines used for the edges should be black with a width of 3, and the face should be grey. The axes (both x and y) should go from 0 to 1. For example, depending on what the random numbers are, the Figure Window might look like Figure 13.29.

9. Using the **patch** function, create a black box with unit dimensions (so there will be 8 vertices and 6 faces). Set the edge color to white so that when you rotate the figure, you can see the edges.

10. Write a script that will create the image seen in Figure 13.30 using a colormap.

11. Write a script that will create the same image, using a three-dimensional true color matrix.

12. Write a script that will generate a 50 by 50 image of pixels. The lower triangular part (including the diagonal) will be all white. The upper triangular part will randomly be either red or green for each element, as shown in Figure 13.31.

13. It is sometimes difficult for the human eye to perceive the brightness of an object correctly. For example, in Figure 13.32, the middle of both images is the same

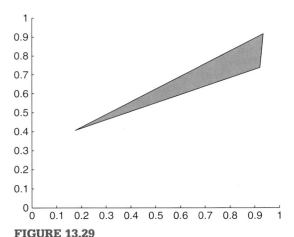

FIGURE 13.29

Patch object with black edge.

FIGURE 13.30

Image displaying four colors using a custom colormap.

FIGURE 13.31

Triangular image of random red and green.

FIGURE 13.32

Depiction of brightness perception.

color—and yet, because of the surrounding colors, the one on the left looks lighter than the one on the right.

Write a script to generate a Figure Window similar to this one. Two 3 × 3 matrices were created. Using the default color map, the middle elements in both were given a value of 12. For the image on the left, all other elements were given a value of 1, and for the image on the right, all other elements were given the value 32. Use **subplot** to display both images side by side (the axes shown here are the defaults).

14. Put a JPEG file in your Current Directory and use **imread** to load it into a matrix. Calculate and print the **mean** separately of the red, green, and blue components in the matrix and also the standard deviation for each.

15. Some image acquisition systems are not very accurate, and the result is ***noisy*** images. To see this effect, put a JPEG file in your Current Directory and use **imread** to load it. Then, create a new image matrix by randomly adding or subtracting a value n to every element in this matrix. Experiment with different values of n. Create a script that will use **subplot** to display both images side by side, using **imshow** instead of **image**.

16. The ***dynamic range*** of an image is the range of colors in the image (the minimum value to the maximum value). Put a JPEG file into your current directory. Read the image into a matrix. Use the built-in functions **min** and **max** to determine the dynamic range, and print the range. Note that if the image is a true color image, the matrix will be three-dimensional, so it will be necessary to nest the functions three times to get the overall minimum and maximum values.

17. A part of an image is represented by an n × n matrix. After performing data compression and then data reconstruction techniques, the resulting matrix has values

that are close to but not exactly equal to the original matrix. For example, the following 4×4 matrix variable *orig_im* represents a small part of a true color image, and *fin_im* represents the matrix after it has undergone data compression and then reconstruction.

```
orig_im =
    156      44     129      87
     18     158     118     102
     80      62     138      78
    155     150     241     105
fin_im =
    153      43     130      92
     16     152     118     102
     73      66     143      75
    152     155     247     114
```

Write a script that will simulate this by creating a square matrix of random integers, each in the range from 0 to 255. It will then modify this to create the new matrix by randomly adding or subtracting a random number (in a relatively small range, say 0 to 10) from every element in the original matrix. Then, calculate the average difference between the two matrices.

18. Put a JPEG file into your Current Directory. Type in the following script, using your own JPEG filename.

```
I1 =imread('xxx.jpg');
[rc h] =size(I1);
Inew(:,:,:) =I1(:,c:-1:1,:);
figure(1)
subplot(2,1,1)
image(I1);
subplot(2,1,2)
image(Inew);
```

Determine what the script does. Put comments into the script to explain it step-by-step. Also, try it using **imshow** instead of **image**.

19. Put two different JPEG files into your current directory. Read both into matrix variables. To superimpose the images, if the matrices are the same size, the elements can simply be added element-by-element. However, if they are not the same size, one method of handling this is to crop the larger matrix to be the same size as the smaller, and then add them. Write a script to do this.

20. Write a function that will create a simple GUI with one static text box near the middle of the Figure Window. Put your name in the string, and make the background color of the text box white.

21. Write a function that will create a GUI with one editable text box near the middle of the Figure Window. Put your name in the string. The GUI should have a call back function that prints the user's string twice, one under the other.

22. Write a function that creates a GUI to calculate the area of a rectangle. It should have edit text boxes for the length and width, and a push button that causes the area to be calculated and printed in a static text box.

23. The Wind Chill Factor (WCF) measures how cold it feels with a given air temperature T (in degrees Fahrenheit) and wind speed (V, in miles per hour). The formula is approximately

$$WCF = 35.7 + 0.6T - 35.7(V^{0.16}) + 0.43T(V^{0.16})$$

Write a GUI function that will display sliders for the temperature and wind speed. The GUI will calculate the WCF for the given values, and display the result in a text box. Choose appropriate minimum and maximum values for the two sliders.

CHAPTER 14

Advanced Mathematics

KEY WORDS

curve fitting
best fit
polynomial
degree
order
data sampling

interpolation
extrapolation
least squares
regression
complex number
real part

imaginary part
purely imaginary
complex conjugate
magnitude
complex plane

CONTENTS

In this chapter, some more advanced mathematics and the related built-in functions in the MATLAB® software will be introduced. In many applications data is sampled, which results in discrete data points. But we often want to fit a curve to the data. *Curve fitting* is finding the curve that *best fits* the data. This chapter will first explore fitting curves, which are simple polynomials, to data. Other topics include complex numbers and a brief introduction to differentiation and integration in calculus.

14.1 FITTING CURVES TO DATA

MATLAB has several curve-fitting functions, and additionally Curve Fitting Toolbox™ has many more of these functions. Some of the simplest curves are polynomials of different degrees, which is what will be described in this section.

14.1.1 Polynomials

Simple curves are *polynomials* of different *degrees*, or *orders*. The degree is the integer of the highest exponent in the expression. For example,

- A straight line is a first order (or degree 1) polynomial of the form $ax + b$, or more explicitly $ax^1 + b$.

- A quadratic is a second order (or degree 2) polynomial of the form $ax^2 + bx + c$.

- A cubic (degree 3) is of the form $ax^3 + bx^2 + cx + d$.

MATLAB represents a polynomial as a row vector of coefficients. For example, the polynomial $x^3 + 2x^2 - 4x + 3$ would be represented by the vector $[1\ 2\ -4\ 3]$.

The polynomial $2x^4 - x^2 + 5$ would be represented by $[2\ 0\ -1\ 0\ 5]$; notice the zero terms for x^3 and x^1.

There are built-in functions **sym2poly** and **poly2sym** that convert from symbolic expressions to polynomial vectors and vice versa, for example:

```
>> myp = [1,2,-4,3];
>> poly2sym(myp)
ans =
x^3+2*x^2-4*x+3
>> mypoly = [2 0 -1 0 5];

>> poly2sym(mypoly)
ans =
2*x^4-x^2+5

>> sym2poly(ans)
ans =

    2     0    -1     0     5
```

The **roots** function in MATLAB can be used to find the roots of an equation represented by a polynomial. For example, for the mathematical function (**Note:** this is a mathematical expression, not MATLAB!),

$$f(x) = 4x^3 - 2x^2 - 8x + 3$$

to solve the equation f(x) = 0:

```
>> roots([4 -2 -8 3])
ans =
   -1.3660
    1.5000
    0.3660
```

The function **polyval** will evaluate a polynomial p at x; the form is **polyval(p,x)**. For example, the polynomial $-2x^2 + x + 4$ is evaluated at x = 3, which yields $-2 * 9 + 3 + 4$, or -11:

```
>> p = [-2 1 4];
>> polyval(p,3)
ans =
    -11
```

The argument x can be a vector, for example:

```
>> polyval(p,1:3)
ans =
     3     -2     -11
>> polyval(p, [5 7])
ans =
    -41 -87
```

14.1.2 Curve Fitting

Data is basically either *discrete* or *continuous*. In many applications, data is *sampled*, for example,

- The temperature recorded every hour

- The speed of a car recorded every one-tenth of a mile

- The mass of a radioactive material recorded every second as it decays

- Audio from a sound wave as it is converted to a digital audio file

This gives data of the form of (x,y) points, which then could be plotted. For example, let's say the temperature was recorded every hour one afternoon from 2 to 6 P.M.; the vectors might be:

```
>> x = 2:6;
>> y = [65 67 72 71 63];
```

and then the plot might look like Figure 14.1.

14.1.3 Interpolation and Extrapolation

In many cases, it is desired to estimate values other than at the sampled data points. For example, we might want to estimate what the temperature was at

FIGURE 14.1

Plot of temperatures sampled every hour.

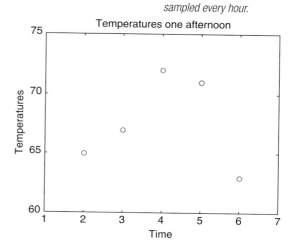

2:30 P.M., or at 1 P.M. *Interpolation* is estimating the values in between recorded data points. Extrapolation is estimating outside the bounds of the recorded data. One way to do this is to fit a curve to the data, and use this for the estimations. *Curve fitting* is finding the curve that "best fits" the data.

Simple curves are polynomials of different degrees as described before. So, curve fitting involves finding the best polynomials to fit the data—for example, for a quadratic polynomial in the form $ax^2 + bx + c$, it means finding the values of a, b, and c that yield the best fit. Finding the best straight line that goes through data would mean finding the values of a and b in the equation $ax + b$.

MATLAB has a function to do this, called **polyfit**. The function **polyfit** finds the coefficients of the polynomial of the specified degree that best fits the data using a least squares algorithm. There are three arguments passed to the function: the vectors that represent the data, and the degree of the desired polynomial. For example, to fit a straight line (degree 1) through the previous data points, the call to the **polyfit** function would be

```
>> polyfit(x,y,1)
ans =
    0.0000 67.6000
```

which says that the best straight line is of the form $0x + 67.6$.

However, from the plot (seen in Figure 14.2), it looks like a quadratic would be a much better fit. The following would create the vectors and then fit a polynomial of degree 2 through the data points, storing the values in a vector called *coefs*.

```
>> x = 2:6;
>> y = [65 67 72 71 63];
>> coefs = polyfit(x,y,2)
coefs =
    -1.8571    14.8571    41.6000
```

This says that MATLAB has determined that the best quadratic that fits these data points is $-1.8571x^2 + 14.8571x + 41.6$. So, the variable *coefs* now stores a vector that represents this polynomial.

FIGURE 14.2

Sampled temperatures with straight line fit.

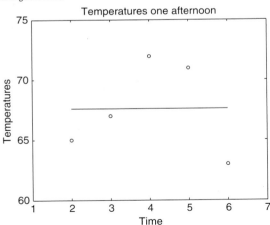

The function **polyval** can then be used to evaluate the polynomial at specified values. For example, we could evaluate at every value in the *x* vector:

```
>> curve = polyval(coefs,x)
curve =
    63.8857      69.4571      71.3143      69.4571      63.8857
```

This results in y values for each point in the *x* vector, and stores them in a vector called *curve*. Putting all this together, the following script called *polytemp* creates the *x* and *y* vectors, fits a second order polynomial through these points, and plots both the points and the curve on the same figure.

polytemp.m

```
%Demonstrates curve fitting
x= 2:6;
y=[65 67 72 71 63];
coefs = polyfit(x,y,2);
curve = polyval(coefs,x);
plot(x,y,'ro',x,curve)
xlabel('Time')
ylabel('Temperatures')
title('Temperatures one afternoon')
axis([1 7 60 75])
```

Calling this results in the plot seen in Figure 14.3. The curve doesn't look very smooth on this plot, but that is because there are only five points in the *x* vector.

To estimate the temperature at different times, **polyval** can be used for discrete x points; it does not have to be used with the entire *x* vector. For example, to interpolate between the given data points and estimate what the temperature was at 2:30 P.M., 2.5 would be used.

```
>> polyval(coefs,2.5)
ans =
    67.1357
```

Also, **polyval** can be used to extrapolate beyond the given data points, for example, to estimate the temperature at 1 P.M.:

```
>> polyval(coefs,1)
ans =
    54.6000
```

The better the curve fit, the more exact these interpolated and extrapolated values will be.

FIGURE 14.3

Sampled temperatures with quadratic curve.

PRACTICE 14.1

To make the curve smoother, modify the script *polytemp* to create a new *x* vector with more points for plotting the curve. Note that the original *x* vector for the data points must remain as is.

Using the **subplot** function, we can loop to show the difference between fitting curves of degrees 1, 2, and 3 to some data. For example the following script will accomplish this for the temperature data:

polytempsubplot.m

```
% Displays curves of degrees 1-3
x = 2:6;
y = [65 67 72 71 63];
morex = linspace(min(x),max(x));
for pd = 1:3
    coefs = polyfit(x,y,pd);
    curve = polyval(coefs,morex);
    subplot(1,3,pd)
    plot(x,y,'ro',morex,curve)
    xlabel('Time')
    ylabel('Temperatures')
    title(sprintf('Degree %d',pd))
    axis([1 7 60 75])
end
```

FIGURE 14.4

Subplot to show temperatures with curves of degrees 1, 2, and 3.

```
>> polytempsubplot
```

creates the Figure Window seen in Figure 14.4.

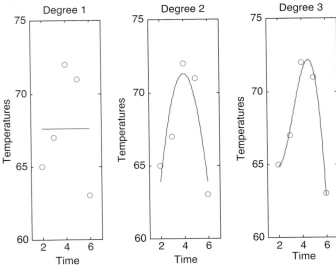

14.1.4 Least Squares

The **polyfit** function uses the *least squares regression* method. To find the equation of the straight line $y = mx + b$ that best fits using a least squares regression, the equations for m and b are:

$$m = \frac{n\sum x_i y_i - \sum x_i \sum y}{n\sum x_i^2 - (\sum x_i)^2}$$

$$b = \bar{y} - m\bar{x}$$

where *n* is the number of points in x and y, all summations are from i = 1 to *n*, and $\bar{y}$ and $\bar{x}$ represent the means of the vectors *y* and *x*. These equations will not be derived here; the derivations can be found in the MATLAB help browser by doing a search for "least squares".

This is implemented as follows in a function *mylinfit* that receives two vectors x and y, and returns the values of m and b. This is the same algorithm used by the **polyfit** function for a degree 1 polynomial, so it returns the same values.

mylinfit.m

```
function [m,b] = mylinfit(x,y)
% least squares regression for a straight line
n = length(x); % Assume y has same length
numerator = n * sum(x .* y) - sum(x)*sum(y);
denom = n * sum(x .^ 2) - (sum(x))^2;
m = numerator/denom;
b = mean(y) - m*mean(x);
```

```
>> x = [-1 1 2];
>> y = [-1 0 3];
>> [m b] = mylinfit(x,y)
m =
    1.2143
b =
    0.1429
>> polyfit(x,y,1)
ans =
    1.2143 -0.1429
```

The least squares fit minimizes the sum of the squares of the differences between the actual data and the data predicted by the line. The "best" straight line in this case has been determined to be y = 1.2143x – 0.1429.

If we did not know that was the best straight line, we might instead guess that the line that best fits the data is the line y = x. The plot is seen in Figure 14.5.

This straight line goes through one of the points, but splits the other two points, in that one is one unit below the line and the other is one above the line. So, it seems as though it fits the data well. However, we will compare this to the line found by **polyfit** and the function *mylinfit*.

FIGURE 14.5

The line y = x and three data points.

Table 14.1 y-Coordinates Predicted by Line y = x

x	Data y	Predicted y	Difference
−1	−1	−1	0
1	0	1	−1
2	3	2	1

Table 14.1 shows the x-coordinates, y-coordinates of the original points, y-coordinates predicted by the line $y = x$, and the differences (data – predicted).

The sum of the differences squared is 0 + 1 + 1, or 2.

According to the least squares algorithm, however, the values using the line $y = 1.2143x – 0.1429$ are shown in Table 14.2.

The sum of the squares of these differences is 1.7857, which is better than (a smaller number than) the sum of the squares of the differences obtained for the earlier straight line. In fact, **polyfit** minimizes the sum of the squares.

MATLAB has another related function, **interp1**, that does a table look-up to interpolate or extrapolate. There are several ways to call this function (using **help** describes them). The default method that is used is 'linear', which gives a linear interpolation. For example, for the previous time and temperature vectors:

```
>> x=2:6;
>> y=[65 67 72 71 63];
```

The **interp1** function could be used to interpolate between the points, for example:

```
>> interp1(x,y,3.5)
ans =
      69.5000

>> interp1(x,y,2.5)
ans =
      66
```

Table 14.2 y-Coordinates Predicted by Least Squares Regression

x	Data y	Predicted y	Difference
−1	−1	−1.3571	0.3571
1	0	1.0714	−1.0714
2	3	2.2857	0.7143

To extrapolate using the linear interpolation method which is the default, the strings 'linear' and 'extrap' must also be passed.

```
>> interp1(x,y,1,'linear','extrap')
ans =
     63
>> interp1(x,y,7,'linear','extrap')
ans =
     55
```

14.2 COMPLEX NUMBERS

A *complex number* is generally written in the form

z = a + bi

where a is called the *real part* of the number z, b is the *imaginary part* of z, and i is $\sqrt{-1}$. This is the way mathematicians usually write a complex number; in engineering it is often written as a + bj, where j is $\sqrt{-1}$. A complex number is *purely imaginary* if it is of the form z = bi (in other words, if a is 0).

We have seen that in MATLAB both **i** and **j** are built-in functions that return $\sqrt{-1}$ (so, they can be thought of as built-in constants). Complex numbers can be created using **i** or **j**, for example, 5 + 2i or 3 – 4j. The multiplication operator is not required between the value of the imaginary part and the constant **i** or **j**.

QUICK QUESTION!

Is the value of the expression 3i the same as 3 * i?
Answer: It depends on whether *i* has been used as a variable name or not. If *i* has been used as a variable (for example, an iterator variable in a **for** loop), then the expression 3 * *i* will use the defined value for the variable, and the result will not be a complex number. Therefore, it is a good idea when working with complex numbers to use 1**i** or 1**j** rather than just **i** or **j**. The expressions 1**i** and 1**j** always result in a complex number, regardless of whether *i* or *j* have been used as a variable.

```
>> i = 5;
>> i
i =
     5
>> 1i
ans =

     0 + 1.0000i
```

MATLAB also has a function **complex** that will return a complex number. It receives two numbers, the real and imaginary parts in that order, or just one number, which would be the real part (so the imaginary part would be 0). Here are some examples of creating complex numbers in MATLAB:

```
>> z1 = 4 + 2i
z1 =
   4.0000 + 2.0000i

>> z2 = sqrt(-5)
z2 =
   0 + 2.2361i

>> z3 = complex(3,-3)
z3 =
   3.0000 - 3.0000i

>> z4 = 2 + 3j
z4 =
   2.0000 + 3.0000i

>> z5 = (-4) ^ (1/2)
ans =
   0.0000 + 2.0000i

>> myz = input('Enter a complex number: ')
Enter a complex number: 3 + 4i
myz =
   3.0000 + 4.0000i
```

Notice that even when **j** is used in an expression, **i** is used in the result. MATLAB shows the type of the variables created here in the Workspace Window (or using **whos**) as **double (complex)**. MATLAB has functions **real** and **imag** that return the real and imaginary parts of complex numbers.

```
>> real(z1)
ans =
   4

>> imag(z3)
ans =
   -3
```

To print an imaginary number, the **disp** function will display both parts automatically:

```
>> disp(z1)
   4.0000 + 2.0000i
```

The **fprintf** function will print only the real part unless both parts are printed separately:

```
>> fprintf('%f\n', z1)
   4.000000
```

```
>> fprintf('%f %f\n', real(z1), imag(z1))
4.000000 2.000000

>> fprintf('%f + %fi\n', real(z1), imag(z1))
4.000000 + 2.000000i
```

The function **isreal** returns 1 for logical true if there is no imaginary part of the argument, or 0 for false if the argument does have an imaginary part (even if it is 0). For example,

```
>> isreal(z1)
ans =
    0

>> z5 = complex(3)
z5 =
    3

>> isreal(z5)
ans =
    0

>> isreal(3.3)
ans =
    1
```

For the variable $z5$, even though it shows the answer as 3, it is really stored as $3 + 0i$, and that is how it is displayed in the Workspace Window. Therefore, **isreal** returns logical false since it is stored as a complex number.

14.2.1 Equality for Complex Numbers

Two complex numbers are equal to each other if both their real parts and imaginary parts are equal. In MATLAB, the equality operator can be used.

```
>> z1 == z2
ans =
    0

>> complex(0,4) == sqrt(-16)
ans =
    1
```

14.2.2 Adding and Subtracting Complex Numbers

For two complex numbers $z1 = a + bi$ and $z2 = c + di$,

$$z1 + z2 = (a + c) + (b + d)i$$

$$z1 - z2 = (a - c) + (b - d)i$$

As an example, we will write a function in MATLAB to add two complex numbers together and return the resulting complex number.

The Programming Concept

In most cases, to add two complex numbers together you would have to separate the real and imaginary parts, and add them to return your result.

addcomp.m

```
function outc = addcomp(z1, z2)
% Adds two complex numbers and returns the result
% Adds the real and imaginary parts separately
realpart = real(z1) + real(z2);
imagpart = imag(z1) + imag(z2);
outc = realpart + imagpart * i;
```

```
>> addcomp(3+4i, 2-3j)
ans =
    5.0000 + 1.0000i
```

The Efficient Method

MATLAB will automatically do this in order to add two complex numbers together (or subtract).

addcomp2.m

```
function outc = addcomp2(z1,z2)
% Adds two complex numbers and returns the result
outc = z1 + z2;
```

```
>> addcomp2(3+4i, 2-3j)
ans =
    5.0000 + 1.0000i
```

14.2.3 Multiplying Complex Numbers

For two complex numbers $z1 = a + bi$ and $z2 = c + di$,

$$
\begin{aligned}
z1 * z2 &= (a + bi) * (c + di) \\
&= a*c + a*di + c*bi + bi*di \\
&= a*c + a*di + c*bi - b*d \\
&= (a*c - b*d) + (a*d + c*b)i
\end{aligned}
$$

For example, for

```
z1 = 3 + 4i
z2 = 1 - 2i
z1 * z2 = (3*1 - -8) + (3*-2 + 4*1)i = 11 -2i
```

This is, of course, automatic in MATLAB:

```
>> z1*z2
ans =
   11.0000 - 2.0000i
```

14.2.4 Complex Conjugate and Absolute Value

The *complex conjugate* of a complex number $z = a+bi$ is $\bar{z} = a-bi$. The *magnitude*, or *absolute value* of a complex number z is $|z| = \sqrt{a^2+b^2}$. In MATLAB, there is a built-in function **conj** for the complex conjugate, and the **abs** function returns the absolute value.

```
>> z1 = 3 + 4i
z1 =
   3.0000 + 4.0000i
>> conj(z1)
ans =
   3.0000 - 4.0000i
>> abs(z1)
ans =
   5
```

14.2.5 Complex Equations Represented as Polynomials

We have seen that MATLAB represents polynomials as a row vector of coefficients; this can be used when the expressions or equations involve complex numbers, also. For example, the polynomial $z^2 + z - 3 + 2i$ would be represented by the vector [1 1 -3 + 2i]. The **roots** function in MATLAB can be used to find the roots of an equation represented by a polynomial. For example, to solve the equation $z^2 + z - 3 + 2i = 0$:

```
>> roots([1 1 -3+2i])
ans =

   -2.3796 + 0.5320i
   1.3796 - 0.5320i
```

The **polyval** function can also be used with this polynomial; for example,

```
>> cp = [1 1 -3+2i]
cp =
```

```
   1.0000    1.0000    -3.0000 + 2.0000i
>> polyval(cp,3)
ans =
      9.0000 + 2.0000i
```

14.2.6 Polar Form

Any complex number $z = a + bi$ can be thought of as a point (a,b) or vector in a *complex plane* in which the horizontal axis is the real part of z, and the vertical axis is the imaginary part of z. So, a and b are the Cartesian or rectangular coordinates. Since a vector can be represented by either its rectangular or polar coordinates, a complex number can also be given by its polar coordinates r and θ, where r is the magnitude of the vector and θ is an angle.

To convert from the polar coordinates to the rectangular coordinates:

```
a = r cos θ
b = r sin θ
```

To convert from the rectangular to polar coordinates:

$$r = |z| = \sqrt{a^2 + b^2}$$

$$\theta = \arctan\left(\frac{b}{a}\right)$$

So, a complex number $z = a + bi$ can be written as `r cos θ + (r sin θ)i`, or

```
z = r (cos θ + i sin θ)
```

Since $e^{i\theta} = \cos\theta + i\sin\theta$, a complex number can also be written as $z = re^{i\theta}$. In MATLAB, r can be found using the **abs** function, and there is a special built-in function to find θ, called **angle**.

```
>> z1 = 3 + 4i;
r = abs(z1)
r =
     5

>> theta = angle(z1)
theta =
     0.9273

>> r*exp(i*theta)
ans =
     3.0000 + 4.0000i
```

14.2.7 Plotting

There are several methods that commonly are used for plotting complex data:

- Plot the real parts versus the imaginary parts using **plot**.

- Plot only the real parts using **plot**.

- Plot the real and the imaginary parts in one figure with a legend, using **plot**.

- Plot the magnitude and angle using **polar**.

Using the **plot** function with a single complex number or a vector of complex numbers will result in plotting the real parts versus the imaginary parts; for example, **plot(z)** is the same as **plot(real(z), imag(z))**. For example, for the complex number z1 = 3 + 4i, this will plot the point (3,4) (using a large asterisk so we can see it!) as seen in Figure 14.6.

```
>> z1 = 3 + 4i;
>> plot(z1,'*', 'MarkerSize', 12)
>> xlabel('Real part')
>> ylabel('Imaginary part')
>> title('Complex number')
```

14.3 CALCULUS: INTEGRATION AND DIFFERENTIATION

The integral of a function f(x) between the limits given by x = a and x = b is written as

$$\int_a^b f(x)dx$$

and is defined as the area under the curve f(x) from a to b, as long as the function is above the x-axis. Numerical integration techniques involve approximating this.

14.3.1 Trapezoidal Rule

One simple method of approximating the area under a curve is to draw a straight line from f(a) to f(b) and calculate the area of the resulting trapezoid as

$$(b-a)\frac{f(a)+f(b)}{2}$$

In MATLAB, this could be implemented as a function.

PRACTICE 14.2

Create the following complex variables:

```
c1 = complex(0,2);
c2 = 3 + 2i;
c3 = sqrt(-4);
```

Then, do the following:

- Get the real and imaginary parts of c_2.

- Print the value of c_1 using **disp**.

- Print the value of c_2 in the form a + bi.

- Determine whether any of the variables are equal to each other.

- Subtract c_2 from c_1.

- Multiply c_2 times c_3.

- Get the complex conjugate and magnitude of c_2.

- Put c_1 in polar form.

- Plot the real part versus the imaginary part for c_2.

FIGURE 14.6

Plot of complex number.

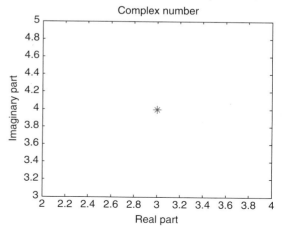

The Programming Concept

Here is a function to which the function handle and limits a and b are passed:

trapint.m

```
function int = trapint(fnh, a, b)
% approximates area under a curve using a
% trapezoid
int = (b-a) * (fnh(a) + fnh(b))/2;
```

To call it, for example, for the function f(x) = 3x² – 1, an anonymous function is defined and its handle is passed to the **trapint** function.

```
>> f = @ (x) 3 .* x .^ 2 - 1;
approxint = trapint(f, 2, 4)
approxint =
     58
```

The Efficient Method

MATLAB has a built-in function **trapz** that will implement the trapezoidal rule. Vectors with the values of x and y = f(x) are passed to it. For example, using the anonymous function just defined:

```
>> x = [2 4];
>> y = f(x);
>> trapz(x,y)
ans =
     58
```

An improvement on this is to divide the range from a to b into n intervals, apply the trapezoidal rule to each interval, and sum them. For example, for the preceding, if there are two intervals, you would draw a straight line from f(a) to f((a + b)/2), and then from f((a + b)/2) to f(b).

The Programming Concept

Here is a modification of the previous function to which the function handle, limits, and the number of intervals are passed:

trapintn.m

```
function intsum = trapintn(fnh, lowrange,highrange, n)
% implements trapezoidal rule using n intervals
intsum = 0;
increm = (highrange – lowrange)/n;
```

(*Continued*)

```
for a = lowrange: increm : highrange - increm
    b = a + increm;
    intsum = intsum + (b-a) * (fnh(a) + fnh(b))/2;
end
```

For example, this approximates the integral of the function f given earlier with two intervals:

```
>> trapintn(f,2,4,2)
ans =
    55
```

The Efficient Method

To use the built-in function **trapz** to accomplish the same thing, the x vector is created with the values 2, 3, and 4:

```
>> x = 2:4;
>> y = f(x)
>> trapz(x,y)
ans =
    55
```

In these examples, straight lines that are first-order polynomials were used. Other methods involve higher-order polynomials. The built-in function **quad** uses Simpson's method of accomplishing this. Three arguments normally are passed to it: the handle of the function, and the limits a and b. For example, for the previous function:

```
>> quad(f,2,4)
ans =
    54
```

14.3.2 Differentiation

The derivative of a function y = f(x) is written as $\frac{dy}{dx} f(x)$ or f′(x) and is defined as the rate of change of the dependent variable y with respect to x. The derivative is the slope of the line tangent to the function at a given point.

MATLAB has a function **polyder**, which will find the derivative of a polynomial. For example, for the polynomial $x^3 + 2x^2 - 4x + 3$, which would be represented by the vector [1 2 − 4 3], the derivative is found by:

```
>> origp = [1 2 -4 3];
>> diffp = polyder(origp)
diffp =
    3    4    -4
```

which shows that the derivative is the polynomial $3x^2 + 4x - 4$. The function **polyval** can then be used to find the derivative for certain values of x; for example for x = 1, 2, and 3:

```
>> polyval(diffp, 1:3)
ans =
     3      16      35
```

The derivative can be written as the limit

$$f'(x) = \lim_{h \to 0} \frac{f(x+h) - f(x)}{h}$$

and can be approximated by a difference equation.

MATLAB has a built-in function, **diff**, which returns the differences between consecutive elements in a vector. For example,

```
>> diff([4 7 15 32])
ans =
     3      8      17
```

For a function y = f(x) where x is a vector, the values of f'(x) can be approximated as **diff(y)** divided by **diff(x)**. For example, the previous equation can be written as an anonymous function.

```
>> f = @ (x) x .^ 3 + 2 .* x .^ 2 - 4 .* x + 3;
>> x = 1:3;
>> y = f(x)
y =
     2      11      36
>> diff(y)
ans =
     9 25
>> diff(x)
ans =
     1 1
>> diff(y) ./ diff(x)
ans =
     9      25
```

14.3.3 Calculus in Symbolic Math Toolbox

There are several functions in Symbolic Math Toolbox™ to perform calculus operations symbolically; for example, **diff** to differentiate and **int** to integrate. To find out about the **int** function, for example, from the Command Window:

```
>> help sym/int
```

For example, to find the indefinite integral of the function $f(x) = 3x^2 - 1$:

```
>> syms x
>> int(3*x^2 - 1)
ans =
x^3-x
```

Instead, to find the definite integral of this function from x = 2 to x = 4:

```
>> int(3*x^2 - 1, 2, 4)
ans =
54
```

Limits can be found using the **limit** function; for example, for the difference equation described previously:

```
>> syms x h
>> f
f =
        @(x) x .^3 + 2 .*x.^2 - 4 .* x + 3
>> limit((f(x+h)-f(x))/h,h,0)
ans =
3*x^2-4+4*x
```

To differentiate, instead of the anonymous function we write it symbolically:

```
>> syms x f
>> f = x^3 + 2*x^2 - 4*x + 3
f =
x^3+2*x^2-4*x+3

>> diff(f)
ans =
3*x^2-4+4*x
```

 PRACTICE 14.3

For the function $2x^2 - 5x + 3$:

- Find the indefinite integral of the function.
- Find the definite integral of the function from x = 2 to x = 5.
- Approximate the area under the curve from x = 2 to x = 5.
- Find its derivative.
- Approximate the derivative.

SUMMARY

Common Pitfalls .

- Forgetting that the **fprintf** function by default prints only the real part of a complex number
- Extrapolating too far away from the data set

Programming Style Guidelines

- The better the curve fit, the more exact interpolated and extrapolated values will be.

MATLAB Functions and Commands			
sym2poly	interp1	conj	diff
poly2sym	complex	polar	int
roots	real	trapz	limit
polyval	imag	quad	
polyfit	isreal	polyder	

Exercises

1. Express the following polynomials as row vectors of coefficients:

   $2x^3 - 3x^2 + x + 5$
   $3x^4 + x^2 + 2x - 4$

2. Find the roots of the equation $f(x) = 0$ for the following function. Also, create x and y vectors and plot this function in the range from –3 to 3 in order to visualize the solution.

   $f(x) = 3x^2 - 2x - 5$

3. Evaluate the polynomial expression $3x^3 + 4x^2 + 2x - 2$ at $x = 4$, $x = 6$, and $x = 8$.

4. Sometimes the roots of polynomial equations are complex numbers. For example, create the polynomial row vector variable *pol*:

   >> pol = [3 6 5];

   Use the **roots** function to find the roots. Also, use **ezplot(poly2sym(pol))** to see a plot. Then, change the last number in *pol* from 5 to –7 and again find the roots and view the plot.

5. Create a vector x that contains the integers 1 through 20. Create a vector y that stores 20 random integers, each in the range from –2 to +2. Fit a straight line through these points. Plot the data points and the straight line on the same graph.

6. The compliance or elasticity of the lung is defined as:

   $$Compliance = \frac{\Delta Volume}{\Delta Pressure}$$

   In a biomedical engineering physiology lab, a spirometer was used to measure the volume of the lungs at a given pressure, which was measured by a pressure transducer. The following data was collected:

Pressure	Volume
0 cmH20	1.750 L
5 cmH20	2.500 L
10 cmH20	3.750 L
15 cmH20	4.000 L
20 cmH20	4.750 L

Write a script that creates vectors to represent this data. Next, the script will find the straight line that best fits this data, and plots the data as 'o's and also the straight line on the same graph. The slope of this line is the actual compliance of the lung and chest wall. Label your axes and put a title on it.

7. The voltage in a circuit is determined at various times, as follows:

```
time:         1      2      3      4      5      6      7
voltage:    1.1    1.9    3.3    3.4    3.1    3.3    7.1
```

Fit a straight line through the data points, and then plot this line along with the sample voltages. According to your straight line, determine at what time the voltage would be 5.

8. Write a script that will generate a vector of 10 random integers, each in the range from 0 to 100. If the integers are evenly distributed in this range, then when arranged in order from lowest to highest, they should fall on a straight line. To test this, fit a straight line through the points and plot both the points and the line with a legend. For example, when tested, the random integers might be

```
95    23    61    49    89    76    46    2    82    44
```

and the plot might look like the one in Figure 14.7.

9. Write a function that will receive data points in the form of x and y vectors. If the lengths of the vectors are not the same, then they can't represent data points so an error message should be printed. Otherwise, the function will fit a polynomial of a random degree through the points, and will plot the points and the resulting curve with a title specifying the degree of the polynomial. The degree of the polynomial must be less than the number of data points, n, so the function must generate a random integer in the range from 1 to n–1 for the polynomial degree.

10. Temperature readings were done every hour (starting at 1 P.M., but the end time could vary) and stored in a vector called *readings*. Write a function called *halffit* that receives this vector as an argument and uses a quadratic interpolation (second order) to determine what the temperature was every half hour between the actual recorded temperatures. The function then plots, on one graph, the original temperature readings (using a 'o' for the points), the interpolated temperatures at the half hours (using a '+' for these points), and the quadratic curve that was used for the interpolation. Put a legend on the graph to distinguish them. The number of hours that was used for the original vector may not be assumed. For example, the function might be called as follows:

```
>> readings = [33, 40, 42, 41, 39, 32];
>> halffit(readings)
```

The Figure Window would look like Figure 14.8.

FIGURE 14.7
Straight line curve fit to random integers.

FIGURE 14.8

Temperatures interpolated every half hour.

FIGURE 14.9

Subplot to demonstrate curves of increasing degrees.

11. Vectors *x* and *y* have been created to represent x and y points. The vectors have the same length (let's call this *n*). Write a function called *fitsubs* that receives these vectors and graphically displays the difference between fitting polynomials of degree 1, 2, 3,...n–1 to the data. For example, *if* the data is as shown here, the Figure Window would look like Figure 14.9.

```
>> x = 1:4;
>> y = [23 35 46 39];
>> fitsubs(x,y)
```

12. Create vectors for four points. Fit a straight line through the points, and also a quadratic. Plot both of these, and the points, on one figure with legends.

13. The temperature (in degrees Fahrenheit) was recorded every three hours for a day at a particular location. Using a 24-hour clock where midnight is 0, for example, the data might be:

```
Time:    0     3      6      9     12     15     18     21
Temp:   55.5  52.4   52.6   55.7  75.6   77.7   70.3   66.6
```

   - Create vectors for the data.
   - Plot the data.
   - Find a curve that fits the data.
   - At what time(s) was it 60 degrees? 65 degrees?

14. The distance (in miles) and speed of a car (in miles per hour) are measured at several points along a highway and are to be stored in a file and then read into a variable called *tripdata*. For example, *tripdata* might contain:

```
  1        44
 10        45
 50        65
100        60
150        55
```

It may be assumed that there are two columns; the first is the distance, and the second is the speed. It may not be assumed that the number of rows is known. The algorithm is:

- Create the data file and load it into a matrix variable, then separate the data into two vectors.
- Fit a straight line to the data.

Plot the data points and the line on the same graph, with appropriate labels on the axes (not just x and y!).

15. Write a function that will receive x and y vectors representing data points. The function will create, in one Figure Window, a plot showing these data points as circles and also in the top part a second-order polynomial that best fits these points and on the bottom a third-order polynomial. The top plot will have a line width of 3 and will be a gray color. The bottom plot will be blue, and have a line width of 2. For example, the Figure Window might look like Figure 14.10. The axes are the defaults. Note that changing the line width also changes the size of the circles for the data points. You do not need to use a loop.

FIGURE 14.10

Subplot of second- and third-order polynomials with different line properties.

16. The depth of snow in inches has been measured in a very cold location every week since the snow began accumulating. At this point, the season has changed and it is getting warmer so the pile of snow is beginning to recede but it hasn't all gone away yet. The depths that have been recorded every week are stored in a file called 'snowd.dat'. For example, it might contain the following:

    8   20   31   42   55   65   77   88   95   97   89   72   68   53   44

    Write a script that will predict in which week the snow will be totally gone by fitting a quadratic curve through the data points. This will be called the "snow gone week number" and will be rounded up. For example, if the data is as shown, the snow would be gone by week number 18. The script will produce a plot in the format shown in Figure 14.11, showing the original data points from the file and also the curve (from week 1 through the snow gone week). The snow gone week number will also be printed in the title. The x-axis should range from 0 to the snow gone week number, and the y-axis from 0 to the maximum snow accumulation.

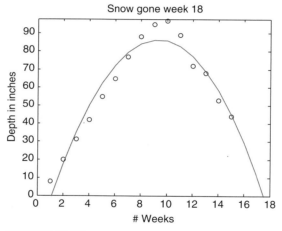

FIGURE 14.11

Prediction of snow melt.

17. A data file called 'acme.dat' stores for the Acme Products Corporation their costs and sales for each of the four quarters last year. There are four lines in the file, each consisting of the costs for the quarter and then the sales. For example, the file might contain the following:

2.2	4.4
4	3.8
6.5	6.5
11.1	10.5

    Write a script that will load this into a matrix and then separate the costs and sales into vectors. Create a Figure Window that shows bar charts for the costs, sales, and profits for the four quarters. Next, extrapolate to determine what the costs are likely to be in the next quarter (assuming a linear progression).

18. Store the following complex numbers in variables, and print them in the form

    $a + bi$.

    $3-2i$

    $\sqrt{-3}$

19. Create the following complex variables:

    ```
    c1 = 2 −4i;
    c2 = 5+3i;
    ```

    Perform the following operations on them:

    - Add them.
    - Multiply them.
    - Get the complex conjugate and magnitude of each.
    - Put them in polar form.

20. Represent the expression $z^3 - 2z^2 + 3 - 5i$ as a row vector of coefficients, and store this in a variable *compoly*. Use the **roots** function to solve $z^3 - 2z^2 + 3 - 5i = 0$. Also, find the value of *compoly* when $z = 2$ using **polyval**.

21. Determine how to use the **polar** function to plot the magnitude and angle of a complex number in polar form.

22. The real parts and imaginary parts of complex numbers are stored in separate variables, for example:

    ```
    >> rp = [1.1 3 6];
    >> ip = [2 0.3 4.9];
    ```

    Determine how to use the **complex** function to combine these separate parts into complex numbers; for example,

    ```
    1.1000 + 2.0000i   3.0000 + 0.3000i   6.0000 + 4.9000i
    ```

23. Using the symbolic function **int**, find the indefinite integral of the function $4x^2 + 3$, and the definite integral of this function from $x = -1$ to $x = 3$. Also, approximate this using the **trapz** function.

24. Use the **quad** function to approximate the area under the curve $4x^2 + 3$ from -1 to 3. First, create an anonymous function and pass its handle to the **quad** function.

25. Use the **polyder** function to find the derivative of $2x^3 - x^2 + 4x - 5$.

Index